THE TWILIGHT OF THE SCIENTIFIC AGE

THE TWILIGHT OF THE SCIENTIFIC AGE

Martín López Corredoira

BrownWalker Press
Boca Raton

The Twilight of the Scientific Age

BrownWalker Press
Boca Raton, Florida • USA
2013

ISBN-10: 1-61233-634-5
ISBN-13: 978-1-61233-634-3

www.brownwalker.com

Frontcover image:
Twilight at the Teide observatory, Canary Islands, Spain.
Photo by Ángel Luis Aldai. Banco de Imágenes Astronómicas
of the Instituto de Astrofísica de Canarias

Library of Congress Cataloging-in-Publication Data

López Corredoira, M. (Martín), 1970-
The twilight of the scientific age / Martín López Corredoira.
    pages cm
Includes bibliographical references and index.
    ISBN 978-1-61233-634-3 (pbk. : alk. paper) -- ISBN 1-61233-634-5
(pbk. : alk. paper)
    1. Science--History. 2. Science--Philosophy. 3. Science--Political
aspects. 4. Regression (Civilization) I. Title.

Q174.8.L65 2013
500--dc23

2012050929

CONTENTS

1 LEITMOTIV

"Yes, yes, I see it; a huge social activity, a powerful civilization, a lot of science, a lot of art, a lot of industry, a lot of morality, and then, when we have filled the world with industrial wonders, with large factories, with paths, with museums, with libraries, we will fall down exhausted near all this, and it will be, for whom? Was man made for science or science made for man?"

– Miguel de Unamuno
Tragic Sense of Life in Men and in Peoples

This quotation reflects quite accurately the main theme of the present book. Read it carefully, twice or thrice, think about it for some minutes, and then begin to read the following pages as a symphony whose *leitmotiv* is Unamuno's assertion. Just a few minutes, or even seconds, may be enough for the reader to realize the most important message that I want to develop, and its connection with the title *The Twilight of the Scientific Age*. The idea is simple: our era of science is declining because our society is becoming saturated with knowledge which does not offer people any sense of their lives. Nevertheless, in spite of the simplicity of this idea, its meaning can be articulated in a much richer way than through one sentence, as in the case of a symphony which develops variations on a folk melody. That will be the aim of this book.

There are several reasons to write about this topic. First of all, because I feel that things are not as they seem, and the apparent success of scientific research in our societies, announced with a lot of ballyhoo by the mass media, does not reflect the real state of things. Also, because the few individuals who talk about the end of science, do so from relativistic or antiscientific points of view, not believing that science really talks about reality, or they relate the scientific twilight to the limits of knowledge. However, there is a lack of works which question the *sense* itself of the pursuit of the truth among present-day thinkers. Of course, there are many humanistic approaches which simply ignore science, but ignoring is not the same as considering its sense or lack of sense. There are many well-prepared scientists or journalists who move in the world of science and consider it in their interactions with the rest of society, but usually they focus too much on the scientific and technical details and do not go deeply enough into existentialist or subjective approaches. A wider vision of both worlds, those of the humanities and science, is necessary to undertake the task. I feel I am able to offer something of this sort, given my experience as both scientist and philosopher. It is not a matter of virtuosity in either scientific knowledge or other areas but a matter of being able to integrate a global view of the fate of our societies. Normally specialists are too focused in their narrow or biased views to offer a global analysis and feeling.

When we talk about the "sense" of something, we cannot undertake a pure analysis in objective terms as in a scientific study. The professional activities on those who dedicate their lives to natural or social sciences usually overlook the fact that, after all, human beings do not move because of *reasons* but because of *emotions*. As psychoanalysis claims, most of our actions are determined by unconscious impulses. And science itself is not an exception: It is made by men whose motivations stem from factors other than a mere pursuit of knowledge. We are not machines, we are not gods; we are just animals, very peculiar animals and very intelligent and curious, that make scientific enterprises work, but subject to multiple internal and external conditions.

Societies as a whole are also sensitive to motivation. As a matter of fact, not all societies developed science. And, as it is known, even civilizations which developed that world-view and that methodology of observing phenomena can decline and lose their interest for continuing the scientific activity. That happened in Western Christian countries in the Middle Ages. Were the Middle Ages a dark age? Possibly, from some intellectual points of view, but it was not the end of civilization. It was an era with plenty of resources to create magnificent things, such as cathedrals. There were means to carry out great advances in many areas. Christianity was not intellectually underdeveloped with respect to Muslim countries, and basic knowledge of Greek science was also present; however, with very few significant exceptions, there was not a great development of sciences in Christian Europe during nearly the ten centuries of the Middle Ages. Why? Maybe because people were not motivated enough to think about nature. Surely, religious context had something to do with this, and the philosophy associated with religion which was ordered to follow faith above all. But possibly this is not the full explanation: The great revival of science in the Renaissance took place within similar religious creeds; also, the Muslim religion was not so different to Christianity and allowed in the Middle Ages a higher development of sciences, declining later when science in Christian countries began to dominate.

In our era, the conditions are very different to the Middle Ages. Nonetheless, in a not very far future, societies embroiled in a lot of survival problems (overpopulation, lack of energy resources, economical crises, global warming and other ecological disasters, wars, plagues, etc.) may begin to see research as an activity that is not profitable enough and may abandon pure science research. At the beginning, people will trust scientists to solve all their problems, as it happens now, but they will realize that science cannot satisfy all those expectations, and that the returns of hyper-millionaire investments are smaller and smaller, nations will reduce more and more the titanic economic efforts necessary to produce some tiny advances in our sciences, to a point where scientists will say that they cannot continue their activity with such small budgets; consequently, the research centres will begin to close, one after another. Is this the prophecy I want to develop in this book? No, I do not want to talk about prophecies. The future is uncertain and what I have described is only one possibility among many others. I want to speak about our present society, and the trends that can be observed now.

Normally, throughout History, thoughts occur in advance of acts. What we are observing around us now are the effects of an ideology which was in some minds many decades or centuries ago. There is a slow inertia in societies which makes them move at the rhythm of impulses that originated some generations back. Geniuses are in advance of their time; what is famous at any moment is representative of a tradition of old, worn-out ideas. Religions gained their maximum power and influence a long time after they were developed: Popes and priests in the Renaissance, embedded in corruption and malpractice, with much less idealism than the conceivers of the religious ideas, were dominant in a time in which the most important creators were pointing to other directions. Today, science and some of its priests occupy an important status in our society, and gargantuan amounts of money support them. A superficial view may lead us to think that we live in the golden age of science but the fact is that the present-day results of science are mostly mean, unimportant, or just technical applications of ideas conceived in the past. Science is living on its private income.

My interest is to lift the curtains behind the stage of science, and see what is going on in the engine room. If we want to ascertain which will be the next performance on the stage, it is better to see the organization from inside rather than just assisting with the show. In any case, I insist, I am not a prophet and it is not my mission to say how the future will be. Also, it is not my mission here to give a report of all the observed trends and ideas around the world of science. What I will offer is my personal view, not necessarily reflecting the views of all conformist and non-conformist present-day thinkers.

The *leitmotiv* is a simple melody. Its harmonization with other melodies and rhythms and the orchestration which integrates all the voices is a more complex thing. As in Wagner's operas, we pursue an infinite melody: A continuous flow where the main melody gets lost among instrumental and human voices. The question of the sense or non-sense of the human endeavour called science must take into account many circumstances. The exhausting of important ideas to explore, the limit of knowledge, is part of the matter. The excess of information is another part. But there are more questions to tackle. The question about the sense of all this stems from those different sources, like a river that takes water from its tributaries, and also from the need for introspective reflection. From time to time, it becomes necessary to go away from the river and contemplate it from the shore. Where does the river go? To the sea, we shall answer. And what for? Is it to achieve Truth? Is it to dominate Nature? What for? For whom? Was man made for science or science made for man?

Thinking about the role of science in present-day society is thinking about the past and the future of humanity. Human beings must question from time to time all their principles and their usual ways of life. There is nothing sacred and untouchable. The missions that science had in the past have been totally accomplished, or almost totally. Now, it is time to reflect anew on our society for the future, not only science but also many other activities or concepts: Art, religions/sects, History, universities, economic systems, political systems, human rights, etc. Very few things are permanent, and all of them are biological, such as taking food and water, sleeping, having sex, etc. All cultural things are subject to change; there is nothing eternal in them. From an anthropological point of view, all the characteristics of our civilization are simple features of the human specie in a given period of time and a given geographical localization.

Certainly, the success of Western culture, with the subsequent annihilation of other cultures, has expanded the geographical location of our civilization to the whole planet, and this might lead us to think that our concepts, such as the so-called human rights, are absolute and universal. A mirage, an illusion! We just live our moment of glory, such as those of many empires which have absorbed great portions of land. The Roman Empire and the Egyptian civilization were greater than us; they lasted longer periods of time, dominating relatively large portions of land for that era. They were perhaps as proud as we are of our Western culture but they eventually declined. Now, it makes no sense to us to bury and embalm the pharaohs under pyramids. Possibly, future civilizations will not see any sense in building huge particle accelerators or telescopes.

You may think that the pharaohs were wrong in their belief that they could preserve life after death, whereas we are right in our scientific truths. I agree. I am not a stupid cultural relativist: Of course, atoms exist and they are constituted by subatomic particles; of course, galaxies and stars exist. But the question is not about the truth of scientific assertions but about the place these truths occupy in our lives as human beings. In the Egyptian civilization or in our civilization, we are moved by our beliefs about what are the high values for our lives. The pharaohs believed that the great architectonic efforts of their people were worth it because that would allow them to be closer to eternity after death, and to show their status on earth too. Scientists believe that dedicating their lives to scrutinizing the laws of nature and making a complete catalogue of all the existing forms of matter, either inert or alive, will bring them closer to something eternal: truth; and make some profit on earth too... But then a question like that of Unamuno arises: "when we have filled the world with industrial wonders, with large factories, with paths, with museums, with libraries, we will fall down exhausted near all this, and it will be, for whom?" Is not it like the child of the tale *The Emperor's New Clothes* that wakes us up from our dreams?

Behind the search for something permanent in our lives, something eternal, something absolute, there is most likely some fear of death. Death is an unavoidable topic if we are going to talk about the sense of some activity for our lives, or the sense of life itself, because precisely our certainty of the finiteness—and indeed very short compared to our aspirations—of our lives pricks our need to search for a sense. We waste our time: we will never find any sense in terms of eternity, but culture is fed mostly because of these aspirations, so the belief is not a bad business at all. Indeed, culture might be understood as the attempts of a civilization to alleviate the tension of the uncertainty which produces our certainty that we are going to die. From this psychological point of view, science is just one of the performances of this tension on stage among many possibilities.

History shows us many dawns and twilights in the different facets of human beings. Looking at the past we can date and understand the reasons for the birth of science. We do not know when its twilight will occur, but the reasons for it are already in the air: after a very hot summer always come the season for the drop of leaves.

1.1 Who has written this book and in which circumstances?

Any book is written by a person or a team of people. It is not something sent by Heaven. And knowing who has written that book is usually an important reference for the reader. Usually, introductions are added to the books, with a description of the author; this is very frequent when it is a classical text and the author is long since dead. For the present book, I prefer to write this introduction myself rather than letting other people to do it. Whether it is read by my contemporaries or by generations after mine, I consider that it is appropriate. I consider that an author must be explicit about all the contents to be transmitted to the reader, and should not wait for somebody else to interpret the messages in terms of biographical aspects. In my opinion, those persons dedicated to writing introductions of books of other more important or classical authors should become jobless. Culture should not be a milk cow on which to grow fat, giving employment to some individuals who cannot produce original ideas and have to make their business with the ideas of others. Whether this book becomes a classic text is difficult, but if it does, the fact that it will not be used to feed the stomach of some mediocre intellectual will be a satisfaction for me, i.e. it is my will that no introductions or footnotes or complementary explanations trying to explain the contents be added to the present text.

I am a Spaniard, born in December 1970 in the town of Lugo, and I have lived in my country most of my life, except for two years in Switzerland and short stays in many other countries. I consider that my education was sound, very good if we compare it with the present-day educational programmes in my country. Secondary school was, in the '80s in Spain, still a way to receive a good grounding in many disciplines. Later, I studied Physics, with the speciality of Astrophysics, and graduated in 1993. One year before the graduation, I began to get in contact with various institutes of research through a special grant for outstanding students of final-year courses, in Vil lafranca del Castillo (Madrid, Spain). There I remained for two years, working and learning things from various theoretical physicists and astronomers in different fields, and developing a postgraduate work on theoretical cosmology. In 1994, I moved to the *Instituto de Astrofísica de Canarias*, where I carried out my work to earn a Ph.D. in Physics in 1997, on the structure of the Milky Way and stellar populations. I think I hold the record at my institute for a Ph.D. obtained in the shortest time: 2 years and 10 months. This was a totally different work from my previous one on theoretical cosmology; rather it was mainly about observational astronomy, although with some application of statistics, which included diverse calculations. This change of orientation allowed me a wider view of astrophysics, including many subfields (stars, galaxies, cosmology) and many techniques (analytical calculations, computer simulations, statistics, photometrical and spectroscopical observations, analysis of data, etc.). After that, I continued with astrophysics as postdoc for 10 years, most of them in the same institute except of two years in Basel (Switzerland). During this time, I worked mainly in the fields of: Milky Way structure, galactic dynamics, large-scale structure of the universe, quasars, and some aspects of the fundamentals of observational cosmology, publishing more than 50 papers in international refereed journal, most of them as first author. Although most of my papers are quite conventional for a professional astrophysicist, there are many of them (maybe 10-20% of them) with a substantial degree of challenge to established ideas. This brought me some fame among my colleagues for holding unorthodox views. Indeed, most of my works are quite orthodox. And indeed, although I have expressed several times my scepticism about some orthodox ideas, in particular about the Big Bang hypothesis in cosmology, I am not an anti-Big Bang cosmologist, as many people have thought. I do not defend any alternative theory. Scepticism, doubting a dogma, does not mean that a new dogma should be defended to substitute for the first one. The opposite thing of dogmatism is criticism rather than a new kind of dogmatism.

Simultaneously with my scientific activities, I developed other activities related to philosophy. In secondary school, I took good introductory courses on some philosophical problems and the history of philosophy. Unfortunately, these courses are disappearing from the educational programmes, becoming in some cases only optional subjects, but I lived in a time and a country in which even "pure sciences" educational programmes in secondary schools included courses on philosophy. The speed of cultural decay of our societies is quite fast, and a generation is enough to appreciate very significant changes. Even in Germany, perhaps the country with the strongest tradition in modern philosophy, it has disappeared from general educational programmes in secondary schools, with Germany becoming a country in which most of the citizens do not know the great thinkers of their country. Possibly this barbarization is an indirect consequence of the right of the winner of a war (Second World War) to impose its criteria: the democratization of culture. Certainly, the criteria of market competitiveness tend to force the educational system towards a specialization in which scientists only know about their speciality and very little or nothing about other fields. In Spain, this "barbarity of specialism", in the words of the Spanish philosopher of the first half of the twentieth century José Ortega y Gasset, is quite notorious now, but 25 years ago there remained still some sense of the long tradition in humanities in this country. I will talk further about the peculiarity of Spain with respect to other northern European countries in the intellectual tradition, but in later chapters, when I talk about Unamuno. The case is that I was educated in a place and a time in which science was becoming increasing important and many resources were rapidly increasing to open many research institutes, and a time in which a good knowledge of history, literature, philosophy, etc. was given to science-oriented students. This does not mean that I lived in a cradle of culture, but compared to the present generations of teenagers, the level was much higher. Actually, in the time I was in high school, when I was a teenager, I was more devoted to science and did not pay too much attention to the humanities. However, once I began my career as physicist I began to worry about subjects in the humanities, mostly philosophy, and the education I had received previously was very useful in choosing the appropriate classical authors to read.

I realized that philosophy was touching some of the questions I was interested in tackling intellectually, which sciences did not touch. Certainly, my way of thinking was that of a scientist, very close to rationalism and logic. I hated those plays on words, that wordiness typical of discourses in philosophy which take too long to say very few clear things. Nonetheless, I realized soon that among the myriads of mediocre intellectuals with nothing interesting to say, there were some lucid thinkers with clear thoughts and very interesting ideas, mostly among classical authors. Indeed, what I read during my years of self formation when I was 19–23 years old were classical philosophers. This marked my ideas as a philosopher in many senses. First, because my formation as a scientist was at a professional level while most of the philosophers just know about science through popular science journals and books at most; second, because I have followed my own way in the selection of reading matters, away from fashion trends or the lines marked by professors and lecturers of a university; third, because I mostly dedicated my time, with very few exceptions, to reading directly (though translated into Spanish) classical texts rather than the opinions of contemporary philosophers about them. There are many other ways to approach to philosophy, certainly, and I do not judge whether my way is better or worse than others, but I think this way has marked my ideas up until now.

In 1994, I initiated a new period in my formation as a philosopher, assisting with some postgraduate courses in philosophy in several Spanish universities, and then developing a thesis which would lead me, in 2003, to getting a Ph.D. in Philosophy, in the University of Seville. The thesis was about the denial of the freedom of will in relation to the contemporary sciences, mainly biology and physics. A book, in Spanish, would be derived from this thesis with the title *Somos fragmentos de Naturaleza arrastrados por sus leyes* ("We are Fragments of Nature Driven by Its Laws"). This period served me to make contact with some academic habits and to understand some of the contemporary problems which are being discussed by the professional philosophers nowadays, mainly in the philosophy of nature and the philosophy of science although also in other fields. I read many works of contemporary philosophers and thinkers in general, and classical works too. I assisted with some conferences of philosophers and some academic events. In general, my appreciation of the philosophy did not change much, and reinforced my idea that among classical philosophers there is an inexhaustible source of wisdom and creative ideas, while among most of my contemporaries I find mostly wordiness, repetitions of old ideas, or crazy snobbish postmodern stupidities. There are exceptions, of course, but perhaps the effort to find them among the rubbish which is written nowadays is too much and life is too short to waste it with that. So I still continue nowadays with the custom of spending more time with classical philosophers than contemporary ones.

I have published some books and articles about philosophy. I have also made some minor contribution to literature, obtaining prizes in poetry and theatre, but these are secondary activities which I have scarcely developed and they are mostly oriented towards literary descriptions of philosophical ideas. I am not a writer of literature. I am a philosopher who uses sometimes literature as a tool to develop philosophical ideas. In general, philosophy has been for me a way to think about the world and our existence as human beings, and the tools to express these impressions may range from science to poetry.

Philosophizing about science, a relevant work which joins my double experience as scientist and philosopher, led to the issue of a collection of papers in English on the sociology of physics and astronomy written from a critical point of view: *Against the Tide. A Critical Review by Scientists of How Physics and Astronomy Get Done.* There I published a polemical paper which was posted on an internet site consisting of preprints by astrophysicists (*arXiv.org*, its old section "astro-ph") five years beforehand: "What do astrophysics and the world's oldest profession have in common?". This paper contained some ideas which will be developed along this book. It was read in *arXiv.org* by thousands of astrophysicists and I received within few days after the early delivery in 2003 more than 50 e-mails commenting on it. Some of the voices were in disagreement, but in most of them I received a message of the type "this is what I think about science but I have never dared to express". I realized that I was not alone in my ideas, and that indeed my observations about science were very common, although very rarely expressed in written texts.

At the moment in which I am beginning to write the present book, I am 39 years old, and I have dedicated more than 15 years to scientific research. In this year 2010, I take a pause in my research activities to reflect on science, while I wait to get a permanent position as researcher at the *Instituto de Astrofísica de Canarias* (I succeeded, and I have got it). I have learnt many things on science, but I have also learnt many things from my sociological observations. My ideas are mature enough, and my blood is young enough to allow impetuosity. There are many ignorant authors (e.g., amateurs) who write books with a lot of pluck and challenge ("Ignorance is daring" says a Spanish proverb), but they say foolish things. There are many experienced people with a lot of academic knowledge who are very accurate in their statements, but they do not dare to talk about things which are actually important or tackle politically incorrect topics;

they produce boring repetitive texts without ideas. Both things are necessary: having experience/knowledge and having a lot of pluck. I think I have something important to say about that human activity called "science", and I think the moment to show my intellectual approach on it has arrived. This is also a time of transition in the world of science, with things declining very fast, and in my own life as scientist. Forty years is an age at which a scientist has to begin to admit that the capacity for innovation and creativity is slowing down, but the capacity to take profit of the accumulated experience is an advantage.

The reader may suspect that the crisis in science I describe in this book is indeed a reflection of a crisis in my life/career. Many times I have heard comments such as "He/she is critical of the system because he/she could not expand his/her horizons and he/she is frustrated by that". Certainly, I agree that there is always a connection between the psychological state of an author and his/her work. And there are many cases in which it is like in the fable of *The Fox and the Grapes*. The fox tries to catch some grapes but it realizes that they are too high so at the end it disdains them saying that they are not mature enough; the moral is that some men disdain the things which secretly they long for but they know to be unreachable. For me science was not an unreachable thing, I have dedicated with pleasure a very important part of my life to it and I have no problem in continuing to work in it. And I have no disdain for science; rather, I love it. Precisely because I love it, I have to raise my voice to preserve the scientific values against the corruption and decadence spreading nowadays. It is true that I do not occupy a high position within the hierarchy of power within the system. I am more a free-thinker, dedicated to my intellectual activities, than a leader of mega-expensive projects or a scientist for the mass media. Nonetheless, I do not envy those high-status positions, and I do not think I am frustrated for not holding them. My major frustration is not about my own creations but perhaps about the lamentable show I have to contemplate, in which intelligence and creativity are disdained whereas technology and money occupy the privileged position, in which poor science is applauded whereas extraordinary ideas are not even commented on. It is indeed a general frustration about the whole culture in most parts of our world: Capitalism gives all the force to people with money, and ideas are only important insofar as they can generate great amounts of money. It is frustrating to see how unfair and how harmful is a world dominated by these market criteria and in which we cannot do anything to stop it. That is the fatal circumstance of our present, and certainly there is not much we can do to save the world from it, but at least we can complain, and this is what I intend to do in this book.

Acknowledgements: Thanks are given to Alexander Unzicker (Physicist, neuroscientist, Pestolazzi Gimnasium, Munich, Germany), Francisco José Soler Gil (Philosopher, Univ. Seville, Spain), Helge Kragh (Historian of science, Univ. Aarhus, Denmark), and Antonio Aparicio Juan (Physicist, Univ. La Laguna, Tenerife, Spain) for helpful comments and suggestions on a draft of this book, and to Maureen Kincaid Speller (www.writersservices.com) for the English-language editing in the final version of the manuscript.

2 SOME HIGHLIGHTS IN THE HISTORY OF NATURAL SCIENCES

If we want to understand why science is declining, we must first understand how it became so important. We must look at some of its advances to realize how important science has been for our culture, otherwise any attempt to talk about the twilight of the scientific era sounds like an anti-scientific complaint, and that is not my purpose. I want to pass on to the readers of this book a love for science, a passion for logic and for understanding the mechanisms which govern nature. There are exciting elements in all this. Science is not just a subject intended to bore students. Rather, it is a fascinating adventure, even more fascinating than science fiction films or the typical elements that the entertainment industry offers to consumers today.

There is plenty of good literature available on the history of science,[1] and I do not intend to offer a new history here. This would occupy many pages, and would not leave space for the main purpose of this book: the interpretation of history rather than the exposition of its facts. Consequently, I will dedicate only a few pages to illustrating the ideas I want to develop, using various representative examples drawn from the natural sciences, to show how great science was in the past by comparison with the minor science which is produced nowadays.

2.1 Greek science

Science, or the philosophy of nature, emerged in Ancient Greece as an attempt to find rational explanations for natural phenomena, overthrowing the idea of gods controlling the universe, emancipating

---

[1] Some of the general sources on the history of science I have used in the following subsections include: Arana (2001); Asimov (1959); Ordóñez, Navarro & Sánchez Ron (2004); Velázquez Fernández (2007); Wikipedia (hereafter WP), as well as many others which are cited in the text. Note: some paragraphs in this chapter are taken from Wikipedia with very few if any changes with respect to the original source. They will be marked (WP) at their end.

nature from the grip of haughty lords and dark, mysterious forces, demystifying the world and facing truth head on. It was an important step for humanity. Thales of Miletus (7-6th century BC) is one of the first known "physicists", those who wondered about the nature of things, and who tried to explain it in terms of its dynamism, or capacity to be transformed under certain principles. The period between Thales and Socrates (5th century BC) is considered as moment of foundation for all the sciences in the western world. Certainly, there were previous civilizations, such as Egypt or Babylonia, or China or India in the eastern world, which had particular knowledge in some sciences, for instance in astronomy, but the impulse to investigate nature was different this time, separate from religion or practical application. It was the acquisition of knowledge for the sake of knowledge within a fairly naturalistic philosophy.

The importance of science in Greek society for the subsequent development of western civilisation was extremely significant, and rather than talking about a timid birth we may talk about an explosion, such is the effect of these centuries of intellectual endeavour. There are plenty of examples concerning the development of sciences to be found in ancient Greece. Many of them involve mathematics, which I will not consider here because they do not directly address nature, but there are also examples in medicine, physics, biology and astronomy, about which I will speak in the next section. Among other impressive examples, I will discuss Hippocrates' work in medicine, Aristotle's in biology, and the work of Archimedes of Syracuse in physics. These examples will show us the roots of the passion for science.

Hippocrates (5-4th century BC), called the father of western medicine, certainly brought the light of rational philosophy to medicine. The aim of his school was to heal rather than simply to study the human body, and his intention was to exclude divine explanations, such as conflict between gods and men, from the possible causes of an illness, by comparing the symptoms and circumstances of different cases of the same affliction. Not a minor thing! Nowadays people still go to quack doctors or charlatans to solve their health problems; imagine how the situation would have been around twenty-five centuries ago. It is the general philosophy of what a human being is and the naturalistic approach, using properly developed techniques rather than simply begging the gods, that is so important. Some techniques of the school of Hippocrates were also important; for instance, in the treatment of wounds, fractures or dislocations. Sometimes, I think that we have not learnt that many new things since the times of Ancient Greek scientists and philosophers; in many cases we have only developed in more detail the ideas they already had. Nowadays we have teams researching in the neurosciences, for instance, but what they are doing is just making explicit what Hippocrates already stated nearly twenty-five centuries ago: "Men should know that their joys, their pleasures, their laughter and pastimes, their sorrow, their grief, their depressions and laments, stem from the brain and only from the brain".

Aristotle (4th century BC) worked in many scientific fields, and he was, together with Plato, one of the most renowned philosophers of Ancient Greece. He and his disciples made a huge collection of observations and documents on which he drew when writing his treatises on natural history. He worked on a general overview on zoology, in which he carried out a detailed analysis of the parts and the functions of animals, with some early ideas on animal behaviour and animal psychology. Unfortunately, Aristotle's writings on botany and animal anatomy have been lost.

Aristotle's classification systems were based on analogies between different animals and their parts. He had already distinguished between vertebrates and invertebrates, what he called "animals with blood" or "animals without blood" (he did not realize that some invertebrates also produce haemoglobin). Animals with blood were divided into live-bearing (humans and mammals) and egg-bearing (birds and fish), while animals without blood were divided into insects, crustacea (shelled or cephalopods) and testacea (molluscs) (WP).

Aristotle carried out research on the natural history of the Greek island of Lesbos, and nearby areas. In the surrounding seas, he made detailed observations on several types of fish, cephalopods, and other sea life. His description of the hectocotyl arm, possessed by the male of most kinds of cephalopods and modified in various ways to effect the fertilization of the female's eggs, was about two thousand years ahead of its time, and widely disbelieved until its rediscovery in the nineteenth century. Among sea animals, he separated mammals from fish, and he knew that sharks and rays were part of the group of selachians (Singer, 1931; WP).

A good example of Aristotle's scientific method is his description, in the work *On the generation of animals*, of breaking open fertilized chicken eggs at intervals to observe when visible organs were generated. He also gave an accurate description of the four-chambered fore-stomachs of ruminants, and made many other observations which indicate a level of scientific endeavour more common in our era (WP).

Aristotle shows in these works the importance of empirical knowledge and a passion for encyclopaedic knowledge. Science comes from patient observation of nature, patient gathering of information and classification of phenomena. Aristotle's background philosophy was that life derives from a pursuit of final causes (teleology), with graded levels of perfection on a ladder of life (scala naturae), rising from plants up to man. It is a false standpoint from a modern biological perspective, but a view which would dominate our understanding of animals and plants until less than two centuries ago. Aristotle's work was so well done and exerted such a long influence in the history of biology that, even twenty-two centuries later, Darwin could say that naturalists such as Linnaeus were mere apprentices in comparison to Aristotle.

Archimedes of Syracuse (Sicily, at that time part of the Greater Greece; 3rd century BC) was one of the most fascinating examples of how to make science into an amazing adventure. Many frontline discoveries and inventions in mathematics, physics, astronomy and engineering are attributed to him, and there are plenty of anecdotes associated with them which indicate his great enthusiasm. Some of the legends about his name might be myths, or they might be real, but in any case they reflect the importance of the science he produced.

The mechanics of the lever is one of Archimedes' contributions to science. Archimedes was supposed to have said: "Give me a place to stand on, and I will move the earth". The king of Syracuse, Hiero II, thought that this was boasting, so he proposed that Archimedes move something very heavy. Archimedes chose a ship loaded with cargo and passengers. Even when empty, the ship could not be moved by a lot of men pulling on ropes. However Archimedes, using ropes and pulleys (an application of the lever principles) was able to move the ship with only one hand. Impressive! I can imagine how fascinated the king and the other people who witnessed it would have been. This was an important step for humanity. It made us more powerful, and it was not magic; it was science!

Archimedes' principle of hydrostatics is no less important, and it is as true and universal today as it was in Archimedes' times: the upward force experienced by a body immersed in a liquid is equal to the weight of the liquid displaced, which corresponds to a volume of the fluid equal to the volume of the immersed body (in an incompressible fluid). Cultural relativists/constructivists usually claim that science is a product of a culture in an epoch and is not appropriate for other epochs and cultures. Examples like Archimedes' principle of hydrostatics, and there are plenty of them as solid as this, show that these relativist assertions are totally out of place. They are not social constructions, but truths about nature, and science is able to elicit some of these truths.

The most popular anecdote about Archimedes' principle does not discuss the dynamic aspect of the principle, but focuses on the more immediate fact that the volume of the displaced fluid is the volume of the immersed body, a great and useful truth with which to measure volume, and consequently, if we know the weight, to derive densities. The anecdote tells us that King Hiero II asked Archimedes to devise a method to check whether a crown ordered to be made by him was of pure gold or whether silver had been added by a dishonest goldsmith. Archimedes had to solve the problem without damaging the crown so he could not melt it down into a regularly shaped body in order to calculate its density. While he was taking a bath, he noticed that the level of the water in the tub rose as he got in, and realized that this effect could be used to determine the volume of the crown. Archimedes then ran into the streets naked, so excited by his discovery that he had forgotten to dress, crying "Eureka!".[2] This word is still used today to announce a brilliant idea. Nevertheless, the practicality of the method it describes has been called into question, due to the extreme accuracy with which one would have to measure the water displacement (WP). Archimedes may have instead sought a solution that applied his principle of hydrostatics to compare the density of the golden crown to that of pure gold by balancing the crown on a scale with a gold reference sample of the same weight than the crown, then immersing the apparatus in water. If the crown was less dense than gold, it would displace more water due to its larger volume, and thus experience a greater buoyant force than the reference sample. This difference in buoyancy would cause the scale to tip accordingly (WP). Whatever the method was, the stroke of genius involved was magnificent, a feat of human intelligence. By the way, the result was that the density of the crown was lower than the pure gold, indicating that the goldsmith had cheated the king by

---

[2] The Greek word "εὕρηκα" means "I have found it".

mixing gold with some lighter metal; consequently, the goldsmith was executed.

Another outstanding episode attributed to Archimedes was in the defence of Syracuse against the Roman army. According to the legend, he used mirrors acting collectively as a parabolic reflector to burn ships attacking the town, and also applied levers to lift up Roman ships and capsize them. The power of science was undeniable. It is easy to understand, with examples like this, how societies became interested in scientific applications: science allows us to become more powerful and dominate nature, but also makes us more powerful fighting our enemies. Beyond the abstractions of mathematical thoughts in the platonic world of ideas, there is a real connection with our actual lives on earth, and Archimedes was one of the most important characters in ancient history to find useful applications in science. Archimedes would defend Syracuse against the attacks of Roman soldiers for three years, but the Roman army finally conquered the town. A soldier asked him to surrender, but he paid no attention to him because he was thinking at that moment in a scientific problem, and simply said "do not disturb my circles!" referring to some geometrical figures he had plotted. The soldier killed him.

2.2 Heliocentric astronomy

Most people do not find the thought of science as pure knowledge particularly attractive. However, there are some instances when it becomes very important, affecting not only specific subjects but also general world views which change our philosophy, religious beliefs, etc. This is the case with the Copernican revolution. That a creature on the earth is able to understand the position and motions of the planet, despite the appearances to the contrary, may be called properly "intelligence".

Again we begin with Greek science, although this history will extend into the modern era. Astronomy is perhaps the oldest natural science. It is also nowadays something akin to the world's oldest profession, but that is another story.[3] Within western culture, there had already been early discoveries in astronomy, such as the ability to use the constellations for navigation, achieved in the early 6th centu-

---

[3] I refer here ironically to my paper "What do astrophysics and the world's oldest profession have in common?" (López Corredoira, 2008a).

ry BC by Thales of Miletus. Pythagoras and his disciples already knew that the earth was spherical and had decomposed solar motion into two components: a yearly one and a daily one.[4] Philolaus, a follower of Pythagoras in the 5th century BC, proposed a model in which the earth, moon, sun and planets all moved around a central fire; since the earth was much closer to this central fire than the rest of the heaven bodies, the earth would be almost in the centre of the universe. More fully developed mathematical models applied to the planetary motions would come in the 4th century BC. Plato proposed that the seemingly chaotic wandering motions of the planets could be explained by combinations of uniform circular motions centred on a spherical earth. Eudoxus of Cnidus, a disciple of Plato, combined several concentric spheres around the earth for each planet, the Sun and the Moon to explain the retrograde motions of some of them. This idea was improved by Callippus, who added seven further spheres to Eudoxus's original twenty-seven. Aristotle would consider the model of concentric spheres to be more than a mere mathematical description and in fact a physical description of reality.

The story of heliocentric models does not begin with Copernicus but much earlier. Heraclides Ponticus (4th century BC) proposed that the earth rotates on its axis, from east to west, but a fully elaborated heliocentric model would be developed later by the Greek astronomer and mathematician, Aristarchus of Samos (3rd century BC). His hypotheses are that the fixed stars and the sun remain unmoved, that the earth revolves about the sun on the circumference of a circle, the sun lying in the middle of the orbit, and that the sphere of the fixed stars, situated about the same centre as the sun, is so great that the circle in which he supposes the earth to revolve bears such proportion to the distance of the fixed stars as the centre of the sphere bears to its surface (Heath, 1913; WP). Aristarchus thus believed the stars to be very far away, and saw this as the reason why there was no visible parallax, that is, an observed movement of the

---

[4] Even nowadays there are people who think that part of the Copernican heliocentric theory is the idea that the Earth is spherical instead of flat. No! Copernicus has not revolutionized astronomy because of that. This was already assumed among cultivated people. As pointed out by Soler Gil (2008), Stephen Hawking in his famous book *A Brief History of Time* (1988) makes an important blunder when he claims that Copernicus contributed to the elimination of the belief in a flat Earth.

stars relative to each other as the earth moved around the sun. The stars are in fact much farther away than the distance that was generally assumed in ancient times, which is why stellar parallax is only detectable with telescopes (WP). Nonetheless, this did not convince the contemporaries of Aristarchus. The only other astronomer from antiquity who is known by name and who is known to have supported Aristarchus's heliocentric model was Seleucus of Seleucia, a Mesopotamian astronomer who lived a century after Aristarchus (WP).

The Aristotelian view prevailed. In the 2nd century AD in Egypt, Ptolemy made a synthesis of the geocentric model and added some improvements which allowed the determination of the position of the planets with higher accuracy, confirming the observations of scientists such as Hipparchus. One of the main components of the model, the idea of epicycles and deferents, had already been suggested by Apollonius of Perga in the 3rd century BC: Mercury and Venus would move around the sun (epicycle), and the sun would move around the earth (deferent); and the rest of the planets would move around a point without any celestial body (epicycle), and this point would move around the earth (deferent). Apollonius's idea was taken up by Ptolemy. He also used the idea of the eccentric, in which the deferents are circles whose centre is not exactly aligned to the earth but slightly off, an idea also present in Apollonius. Ptolemy also added a further complication to the system with the new idea of equants, according to which the linear velocity of the planet is not constant.

Ptolemy thought his system was a true description of reality rather than a mere instrumentalist approach, as stated in his *Planetary Hypotheses*. It is a great demonstration of the power of mathematics to describe phenomena. Beliefs—that the earth does not move, and that any motion in that perfect world of heavens should be in circles—might change, but the science was good. The Ptolemaic model was a superb attempt to control heavenly phenomena in the ancient era, a scientific cosmological model, but it is also a classical example of *ad hoc* science which would establish a precedent of what should not be done: when your theory does not fit the observations/experiments, modify the theory with multiple patches and add as many complications as you need until you get an agreement. Nowadays, nearly two millennia later, we still build cosmological models using many *ad hoc* elements to meet the qualifications which arise in the gathering of observations: inflation, dark matter, dark energy, etc., and cosmologists are as happy now as Ptolemy was when they are able to claim, "Our model reproduces the observations". One might wonder whether we are repeating the same historical errors or whether the cosmos is so complicated (López Corredoira, 2009a). Are not inflation, dark matter, dark energy, etc. elements in a theory equivalent to the deferents, epicycles, eccentric, equants in Ptolemy's model? We do not yet have the answer, and we will have to wait some decades or even centuries to be sure.

The Ptolemaic system was too complicated. Alfonso X the Wise (1221-1284), King of Castile and other conquered regions (of Spain), said: "If the Lord Almighty had consulted me before embarking upon Creation, I should have recommended something simpler". Certainly, the lack of simplicity was perhaps the main reason to search for other less *ad hoc* hypotheses. There were occasional speculations about heliocentrism in Europe before Copernicus. In Roman Carthage, Martianus Capella (5th century AD) expressed the opinion that the planets Venus and Mercury did not go around the earth but instead circled the sun (Stahl, 1977; WP). During the Late Middle Ages, Bishop Nicole Oresme discussed the possibility that the earth rotated on its axis, while Cardinal Nicholas of Cusa in his *De Docta Ignorantia* asked whether there was any reason to assert that the sun (or any other point) was the centre of the universe (WP). Nicholas wrote that "Thus the fabric of the world (machina mundi) will quasi have its centre everywhere and circumference nowhere".

Away from the Christian world, some scientists worked on ideas which were similar to heliocentrism. There were earlier computational systems that may have implied some form of heliocentricity, notably the model devised by the Indian astronomer, Aryabhata (476-550): here the earth was taken to be spinning on its axis and the periods of the planets were given with respect to the sun. Nilakantha Somayaji (1444–1544) developed a computational system for a partially heliocentric planetary model, in which the planets orbit the sun, which in turn orbits the earth; it incorporated elliptic orbits and the earth's rotation on its axis (WP). In any case, these were methods of computation and fell short of proposing models of the universe. The concept of heliocentrism was also considered to be a philosophical problem rather than a mathematical problem in the medieval Islamic civilization, because of the scientific dominance of the Ptolemaic system. Despite that, several Muslim astronomers also developed computational systems with astronomical parameters compatible with heliocentricity. Abu Ma'shar (9th century) developed a planetary model in which the orbital revolutions of the planets are given as heliocentric revolutions rather than geocentric revolutions, and the only known planetary theory in which this occurs is in the heliocentric theory (van der Waerden, 1987). The Persian scientist Biruni (11th century) suggested that if the earth rotated on its axis this would be consistent with astronomical data. And many others, such as Alhazen (11th century), Fakhr al-Din al-Razi (12th century), Najm al-Dīn al-Qazwīnī al-Kātibī (13th century), Qotb al-Din Shirazi (13th century), Ibn al-Shatir (14th century), the astronomers of Maragha and Samarkand observatories such as Al-Tusi (13th century) and Ali Qushji (15th century), also discussed the possibility of a heliocentric system.

The search for truth in science is a heroic enterprise, and the greater the difficulty in reaching it, the more value it has. In the revolution which bears his name, Nicolaus Copernicus (Poland, 1473-1543) was the great hero, although not the only one, to judging from the number of people before and after him who worked on the same idea. This Polish scientist was indeed not a very revolutionary character and he was very cautious about revealing his heliocentric hypothesis, since he knew that scholars were usually dogmatic and intransigent, but he dared to defend heliocentrism. Copernicus discussed the philosophical implications of his proposed system, elaborated it in full geometrical detail, used selected astronomical observations to derive the parameters of his model, and wrote astronomical tables which enabled one to compute the past and future positions of the stars and planets. In doing so, Copernicus moved heliocentrism from philosophical speculation to predictive geometrical astronomy (WP). Around 1530, he presented his theory in a manuscript and allowed it to circulate among many people. He found people who were enthusiastic about the hypothesis, but also many enemies, like Martin Luther, who allegedly said that Copernicus was a fool who denied the Bible. Around 1540, Rheticus, a disciple of Copernicus, published a book about the Copernican theory. Pope Paul III approved that book and asked that the full manuscript by Copernicus be published. Hence, Copernicus was able to publish his work, *De Revolutionibus Orbium Caeletium*, which was dedicated to the Pope; he also added a strong attack against people who used Bible citations to refute mathematical demonstrations. The book was published in 1543, on the same day that Copernicus died. Maybe it was as well that he did not see that the printed version included a preface added by Osiander, who helped to get the book published, claiming that the Copernican theory was just a mathematical juggle to simplify the calculations of the planetary orbits, so there was no truth in the heliocentric model, while the geocentric model was unquestionable. Consequently, the reaction produced by this book was not that strong, although even so some people were disturbed by it. For instance, some years after the publication of *De Revolutionibus*, John Calvin preached a sermon in which he denounced those who "pervert the course of nature" by saying that "the sun does not move and that it is the earth that revolves and that it turns" (Rosen, 1995).

Certainly, heliocentrism is, together with the theory of evolution, the most representative of the conflicts between religion and science. Giordano Bruno was burned at the stake for many heretical ideas which included the heliocentrism, while Galileo Galilei (Italy, 1564-1642) had serious problems with the Catholic church because of his strong support of heliocentrism. He discovered four satellites orbiting Jupiter, the phases of Venus, the spots of the sun, the craters and mountains of the moon, and all of them were considered to be a challenge to the established Aristotelian idea that the heavens were perfect and the earth was the centre of everything. He carried his telescope to Rome in 1611 to show his discoveries to the representatives of the Church and, although some of them found the discoveries interesting, many did not want to make observations through the telescope. They said that Jupiter's moons were not visible to the naked eye, so they had no utility to human beings, and therefore they could not have been created; if the instrument allowed men to see them, it was an artefact of the instrument, a spotted device, a devil's instrument, according to some of them. As is known, two decades later Galileo was obliged to publicly retract his heliocentric ideas. Before that, in 1609, Johannes Kepler (Germany, 1571-1630) had developed a heliocentric model of the solar system in which all the planets had elliptical orbits. This significantly increased the accuracy of predicting the position of the planets, but Kepler's works were placed on the index of prohibited books and his ideas were not widely disseminated at that time.

The heroic martyr is another figure in science. Abstract theories are important, mathematical juggles are amusing and useful for elements of engineering, but the real essence of science is in its ability to enlighten people, the power of reason to demystify the mysteries of nature, the struggle to achieve truth without resorting to ancient mythologies or superstitions. As we have seen, many people throughout history saw the convenience of a heliocentric system, but most of them did not dare to fight for the idea even though they thought it was the truth. This is the mission of the scientist too, not only to perform calculations or experiments but also to fight like a warrior against the dark forces of obscurantism. To use a slang expression, a scientist must have guts, and intelligence of course; the rest is mere entertainment for technicians.

Once superstitions or false beliefs vanish, as in the case of the geocentric system, research about a subject is no longer so exciting. The fact that we can determine nowadays the positions of the planets and all their interactions with an incredible precision does not add much interest to the original realisation that the earth moves around the sun like the rest of the planets, and not the sun and the planets around the earth. Nonetheless, other important revolutions would come in astronomy, the last one being after the end of the Great Debate (Shapley-Curtis debate) in 1920, with the subsequent development of extragalactic astronomy and the conclusion that the sun is not at the centre of our galaxy, and that the Milky Way was not at any privileged position in the universe but only one among many other galaxies. In my opinion, nothing as important and undeniable as this has been discovered in astronomy since then.

This case is also an example of the extension of the limits of science. "This is not a scientific topic" is a very common observation throughout the history of ideas. Scholars for many centuries thought that heliocentrism vs. geocentrism was a metaphysical question rather than a scientific subject, but they were wrong. Even today, we have to hear from time to time that some topics (determinism, origin of consciousness, love, etc.) are not to be treated in scientific terms. Usually, the attempt to find scientific explanations for almost everything is pejoratively called "scientism". However, now the position of science is different: we may say that science has a dominant position in society and anti-scientific ideas are spread like guerrilla bands among resentful philosophers who miss the metaphysical speculations,[5] or among non-intellectual people. Science has won the

---

[5] Here I mean metaphysical speculations separate from beliefs within science. Of course, scientific principles also have a metaphysical basis. There are also some ideologies which do not derive from any empirical fact in the scientific representation of the world. And there is much metaphysics contained in the scientific view of some difficult problems, such as consciousness, ontological indeterminism, etc. However, we can at least say that

war; whether this is good or bad is debateable, but it has won the war after many battles, and one may wonder whether further battles are necessary since the triumph has been obtained.

2.3   Understanding the mechanics of nature

When Galileo was seventeen years old, he observed, while he was assisting at a mass in a cathedral at Pisa, how the lamps oscillated because of the draught, sometimes in large arcs, at other times in smaller arcs. He used his own pulse to measure the period of time in each oscillation in the different instances, and he observed that the time taken was the same, irrespective of the amplitude of the oscillation. He discovered the law of the pendulum, which he would later demonstrate mathematically, and this forms the basis of pendulum clocks. Examples like this, or the famous legend of Newton, the apple and the discovery of gravity, illustrate how we might find inspiration in science. Human intelligence and the training obtained through long study among books are important, but what is more important in order to ascertain new truths is our curiosity and our observations of nature. Apparently, our present-day educational systems have forgotten this aspect. Certainly, many of the most important discoveries were made away from a university office or a research centre. Science is not a mere abstract exercise for solving academic problems. It is the result of the interaction of nature with human intelligence at the highest level.

Human beings have always felt the need to search for explanations for what happens around us, possibly because of the high capabilities of our brains, which may provide a more complex response to impulses received than the brains of other animals. This necessity was channelled by religions in most primitive societies, together with the fear of death. The importance of science at that time was that there were many phenomena around us whose logic was not understood and, in the era of rationalism and the subsequent

---

science may say something about these topics in a positive way rather than just a speculative one, and no territory is far from a scientific perspective. Everything in our universe is a natural phenomenon, and natural phenomena may be studied in a rational-empirical way; that is the main metaphysical principle of scientific research. There remain mysteries to be solved, but about these nothing can be said with any degree of reliability except through science.

century of enlightenment, science would offer explanations in a positive way, distinct from metaphysical speculations. The period between the seventeenth and nineteenth centuries cannot be repeated. Nowadays, wherever we look, everything has already been explained scientifically and one has to design very expensive experiments to make observations that provide science with new data which may lead to new discoveries. Galileo was not looking for new data from new phenomena to build his science; it was the phenomena which accidentally found Galileo's brain in order to react with it and produce such a wonderful result in Galileo's ideas in physics or astronomy. There is no intention to discover, or a pursuit to find something new; the discovery is just a consequence of the intellectual development of human society in its interaction with nature. It is the anticipated fruit of an intellectually advanced society. This, together with the need of new technologies and applications of science, fed the development of science in that era. Also, society was satisfied with those fruits, obviously with the technical advances but also with knowledge for the sake of knowledge. It was a science close to human experience. Everybody could watch the oscillations of a lamp, or the fall of an apple, and could obtain some sense of control over nature by reading the explanations of Galileo or Newton. This is an important difference with present-day science which, like present-day art, is oriented toward specialists' experiences and totally disconnected from the general worries of ordinary people in society. Certainly, there are many propaganda campaigns in the mass media to attract the interest of people to certain topics within astrophysics (the search of new planets, cosmology), biology (genetics), medicine, etc. Nevertheless, apart from the almost metaphysical questions which have always interested human beings (the origin of the universe, whether we are alone in the universe, etc.), society does not demand explanations of ordinary experiences because they have already been produced, and this makes the purpose of scientific research different. But let's leave this question for later chapters and continue with Galileo and Newton.

In some sense, scientific discoveries are like verses inspired by muses. There is no one formula to produce genius-level science in the same way that there is no one formula to produce good poetry. They come when they come, although of course, intelligence and good preparation is necessary and, as said by the painter Picasso: "...that the inspiration finds you while you are working". The man who worked on nearly half of the important issues in the physics we know nowadays was Isaac Newton (UK, 1643-1727). The admiration which Newton's theories earned him, and the good reputation which science earned, marked the beginning of an era of high regard for science, one which still persists. The English poet Alexander Pope, a contemporary of Newton, expressed it thus in his verses:

"Nature and nature's laws lay hid in night;
God said 'Let Newton be' and all was light".

Possibly this admiration is exaggerated, and not all the ideas attributed to Newton come exclusively from him. His *Philosophiae naturalis principia mathematica* (1687) is indeed a synthesis of many ideas: the kinematics of Galileo, the astronomy of Kepler, the atomism of Gassendi, some ideas on mechanics and motion from Descartes, and others. Conflicts about the paternity of some ideas were also very common in the life of Newton the genius. In the case of his theory of gravitation, Robert Hooke, a many-sided scientist, another member of the Royal Society like Newton, had mentioned his ideas on gravitation in terms of attraction in a letter to Newton and also "that the Attraction always is in a duplicate proportion to the Distance from the Center". This was six or seven years before the publication of the *Principia*; however, Newton did not even make any citation or give any acknowledgement to Hooke, and did not want to share the credit for his achievements. Certainly, Hooke might have had the idea, but without the full mathematical developments and demonstrations that Newton delivered. Moreover, the assumption of an inverse proportion with the square of the distance in gravity was rather common and had been advanced by a number of different people for different reasons since the 1660s (Gal, 2002; WP). Hooke claimed that the idea was his, but Newton did not consider it to be so, refusing to accept that somebody else was the author of the theory of gravitation, and maintaining a long enmity with Hooke.[6] Conflicts like this were common in Newton's life. He was apparently very reluctant to recognize the worth of his contemporaries. A first-

---

[6] Hooke and Newton also had a dispute over the authority of some ideas about optics. Hooke died in 1703, and Newton continued as President of the Royal Society for some years more. For some mysterious reason, at that time the only known portrait of Hooke disappeared; it is thought that Newton was responsible for this. Newton tried to ensure science forgot the name of Robert Hooke, despite his numerous contributions.

class genius, indubitably, but there were also some other less brilliant scholars who also deserved some of the recognition he accrued.

Understanding the mechanics of nature was a dream pursued by human reason during many centuries and, with Newton's system, the great triumph of modern physics was reached:[7] a set of principles and laws which allow us to understand and to calculate motion in terms of kinematical principles and known forces, a triumph also in the understanding of gravitation as well as of celestial mechanics. Still, some mysteries remained, such as the stability of the orbits of the planets, which Newton attributed to a periodic intervention on the part of God. More than a century later, this problem would be solved by Laplace, who would reply to Napoleon Bonaparte when asked about the role of God in it: "I had no need of that hypothesis". From Newton onwards, until the beginning of the twentieth century, the remaining tasks of physics concerning mechanics would lie mainly in applying these laws, or understanding those forces which were still unknown. However, other areas like optics or thermodynamics would necessitate independent developments.

The most important contribution to the understanding of the mechanics of nature and its forces after Newton involved another synthesis of knowledge, this time on the subject of electromagnetism, produced by James Clerk Maxwell (UK, 1831-1879). Again, we have to say that not all the credit should go to Maxwell. In 1861 he brought together equations formulated by Gauss, Ampère and Faraday, four partial differential equations that related the electric and magnetic fields to their sources, charge density and current density; and he added a new term to one of the equations (Ampère's law). This correction was, however, very important. It meant that a changing magnetic field creates an electric field, and a changing electric field creates a magnetic field. Therefore, these equations allowed self-sustaining electromagnetic waves, namely "light", to travel through space. Maxwell had achieved the unification of electricity, magnetism and optics. The unification of electricity and magnetism had already been glimpsed in the science of Oersted and Faraday, but the explanation of light in terms of electromagnetism was something totally new, a discovery of the first order made by Maxwell.

---

[7] The term "physics" is generally used after 1850, instead of "natural philosophy".

It is clear from the two greatest achievements of Newton and Maxwell that a good knowledge of mathematics was very important in order to develop a unified view of the results which many experimentalists had achieved. Nonetheless, we must not forget that physics is an empirical science and, although the credit is many times greater for the mathematician who puts the cherry on the top of the cake, the previous cooks in laboratories and observatories are very important too.

The unification achieved by Maxwell opened the way towards the last dream of the physicist, one which persists to the present day: the unification of all physical phenomena into a set of universal laws. And indeed, we could even say more: nineteenth-century science could relate the different branches of science to each other. Chemical phenomena could be reduced to physical phenomena (see subsect. 2.4), and in the nineteenth and twentieth centuries biological phenomena would be reduced to chemistry (see subsect. 2.7). So, a wider picture of the universe emerged, in which all phenomena were indeed physical, and could be understood in mechanical terms. Optics was reduced to electromagnetism. Thermodynamics could be understood in terms of Newtonian mechanics through statistical physics. Understanding the mechanics of nature would mean understanding how nature worked in all its aspects. That was the major aim of science, a dream pursued for a long time. This view, as seen by Huxley (1895) and many others at the end of the nineteenth century, is called "reductionism". Most present-day professional philosophers are against reductionism. They see it as a major threat to metaphysical speculation, and they prefer to adopt either a non-naturalist position or claim a mysterious emergence of irreducible properties in nature. Most present-day scientist however do not even talk about reductionism but take it as an undeniable fact.

According to some authors (e.g., Horgan, 1996, ch.3; Iradier, 2009), this idea of unity, of finally knowing, is an idea derived from the attributes of divine omnipotence, and possibly motivated by the same impulse that conceived the idea of God. Certainly, whatever its psychological/sociological explanation, there is a component of the mini-god complex in all these great attempts to develop a final theory, and this has prompted society to become focused on the great truth of the theory of everything. Indeed, some optimists, such as Lord Kelvin, Hertz, Simon Newcomb, and Albert A. Michelson, thought at the end of the nineteenth century that the full extent of science had already been reached, and the end of science had arrived.

However, there remained some loose ends, which would lead to some new discoveries in physics at the beginning of the twentieth century. One of the loose ends was implicit in the incompatibility of Maxwell's equations with Newton's mechanics. Maxwell's wave equations only applied in what he believed to be the rest frame of a luminiferous medium called aether. Einstein's theory of special relativity (1905) postulated the absence of any absolute rest frame, dismissed the aether as unnecessary and established the invariance and validity of Maxwell's equations in all inertial frame of reference (WP). Newton's equations would have to be modified, although the modifications are negligible for systems with velocities much lower than light speed. Also, in the twentieth century, general relativity (a new theory of gravitation developed by Einstein), quantum mechanics and the discovery of two new forces in the nuclei of atoms (weak and strong) would make the mechanics of nature somewhat more complex, and thus make it more difficult to achieve a unification. Einstein tried during the last twenty years before his death in 1955 to achieve that unification but failed. It is a problem even now.

2.4   Atoms and the structure of matter

The idea that matter is composed by atoms (indivisible particles of different types) moving in an empty space in a mechanical universe without purpose was proposed by Democritus (5-4th century BC), who took it from his master, Leucippus. This is indeed one of the first reductionist ideas, and shows us once again that the most important philosophical concepts were already present among the Greek thinkers. "Atomism" was the idea that everything around us, even human beings, is just collections of atoms moving in an empty space. Apart from the scientific description, there is an ideology in it: materialism, which would remain one of the most important currents of thought up to the present day.

Democritus's idea was sound, and had some followers during subsequent centuries, the Epicureans, for instance, but it was abandoned in the Middle Age, in western societies, and not recovered until much later. In other cultures, there were also independent proposals of the idea of atomism. In Hindu philosophy, the Vaisheshika (2nd century BC) and Nyaya (2nd century AD) schools developed elaborate theories on how atoms combined into more complex objects (first in pairs, then trios of pairs), but believed the interactions were ultimately driven by the will of God, and that the atoms themselves were otherwise inactive, without physical properties of their own (Teresi, 2003; WP). The speculations of the eighth-century Islamic alchemist, Jabir ibn Hayyan, posited the theory of "corpuscularianism" (Levere, 2001; WP) by which all physical bodies possess an inner and outer layer of minute particles or corpuscles which, unlike atoms, may be divided; this idea inspired the alchemists of later centuries.

The seventeenth century saw a resurgence of atomic theory, primarily through the works of Descartes and Gassendi, although the description of the basic constituents of substances was still very primitive and speculative. In a more concrete manner, the concept of aggregates or units of bonded atoms, i.e. molecules, traces its origins to Robert Boyle's 1661 hypothesis: that the matter is composed of cluster of particles and that chemical change results from the rearrangement of the clusters (WP). Certainly, the present-day idea of "atoms" was not yet realized but these were important steps towards it: the non-continuous nature of matter and the idea that substances are composed of very small (invisible indeed) particles was becoming clearer.

At the beginning of the nineteenth century, John Dalton (UK. 1766-1844) introduced the idea that chemical substances are always composed of some kind of particles, and can combine to form complex structures. In principle, it was thought that these particles were indivisible; they could not be altered or destroyed by chemical means, so they were called "atoms", as in Democritus's proposal. Today we know they are not indivisible; atoms are constituted from other subatomic particles, but the name remained. What Dalton proposed was more than a speculation, as Democritus's or Leucippus's ideas had been. It was supported with scientific arguments drawn from various observations of chemical components. The importance of science and its methods in the search for knowledge is very clear in this example, and with it we can understand why science would transform sterile metaphysical arguments about nature. Dalton used earlier observations to achieve his atomic hypothesis, such as the law of conservation of mass, formulated by Lavoisier in 1789, which stated that the total mass in a chemical reaction remains constant; or the law of definite proportions, proven by Proust in 1799, which states that if a compound is broken down into its constituent elements, then the masses of the constituents will always have the same proportions, regardless of the quantity or source of the original substance (WP); but also his own deductions on the observed law of multiple proportions (for instance, "that the elements of oxygen may combine with a certain portion of nitrous gas or with twice that portion, but with no intermediate quantity", in his words), and that the weights of elements are multiple of the weight of hydrogen.[8] Dalton's theory had some flaws which would be corrected later by other chemists. For instance, he did not realize that with some elements, atoms exist in molecules, e.g. pure oxygen exists as O_2. He also mistakenly believed that the simplest compound between any two elements was comprised of one atom of each, so he thought water was HO, not H_2O.

Another line of research which considered the possibility of matter being composed of very small particles was statistical physics. In 1738, Daniel Bernoulli laid the basis for the kinetic theory of gases, defending that gases consist of great number of molecules moving in all directions, that their impact on a surface causes the gas pressure that we feel, and that what we experience as heat is simply the kinetic energy of their motion (WP). The theory was not immediately accepted at that time. Maxwell and Boltzmann in the second half of the nineteenth century would develop a very satisfactory kinetic theory of gases. Another interesting development, the discovery of Brownian motion, by the botanist Robert Brown, in 1827, who noted the fact that grains of pollen floating in water jiggle constantly, was explained by Einstein in 1905 in terms of water molecules continuously knocking the grains. All these scientific achievements reinforced the idea that matter is composed of molecules which, together with the ideas about molecules propounded by chemists, led the atomic hypothesis to become a solid theory.

At the beginning of the twentieth century some scientists, such as the physicist and philosopher Ernst Mach or the chemist and philosopher Wilhelm Ostwald, were still sceptical about atomic theory. But their doubts would soon be dissipated, mainly thanks to work by Jean Perrin on colloids (1908-1909) which confirmed the values of the Avogadro number and Boltzmann constant. The development of atomic theory is, in my opinion, one of the most important achievements of science and it is one of the symbols of solid knowledge. Richard Feynman, winner of Nobel Prize in Physics, also thought that it was the most important theory in the history

---

[8] This idea, although implicit in Dalton's ideas, was proposed in 1815 by William Prout.

of science. Of course, there are still some postmodern relativist philosophers who dare to claim that atomic hypothesis is not yet proven, and that it is just a mere hypothesis. Yet just the opposite is true; if there is one fact which is not evident (the existence of atoms can be derived only indirectly from the observational phenomena) and yet is proven with absolute certainty, it is atomic theory. And, since everything in nature is matter constituted by atoms, the understanding of this fact is of capital importance. Practically, all sciences are fed by atomic theory.

Subatomic particles were discovered and it was demonstrated that atoms have structure and are not indivisible. In 1897, Thomson discovered the electron and in 1909 Rutherford discovered the nuclei of atoms. The first models of atom structure would be produced in the following years, culminating with the quantum physical model of the atom. These are exciting moments in the history of physics, but few decades would be sufficient to draw out the important knowledge about atoms.

In the last decades, particle physicists have made great efforts to learn more about the structure of matter, but with less spectacular results. They have been distracted by abstract Pythagorean conjectures into a search for mathematical symmetries in nature and into very expensive experiments to reach high energies, trying to emulate the successes of that golden age of particle physics in the first half of the twentieth century. Some results were obtained but they were of secondary importance and worse, even in the twenty-first century, many of them were speculative; the only impressive thing about particle physics nowadays is the gargantuan amount of money which it is able to consume in a short time. The limits of ordinary experience have already long since been surpassed, and we have to go a long way to find new challenges.

2.5 History of the earth

Another important conceptual revolution was the discovery of how old our planet is. Rather than cosmology, it was the science of geology which first showed us about the huge amount of time the earth had already existed before human beings arrived. If astronomy, with its development of galactic and extragalactic astronomy, has made us realize how small is the portion of space in which human beings live, geology and, later, biology and astronomy have made us realize how small is the portion of time in which we have lived.

Still, in the early seventeenth century, scientists and philosophers had no any better idea of the age of the earth than that given in the Bible, around six thousand years. In the eighteenth century, however, it was spoken about in terms of millions of years. Geologists such as Abraham Gottlob Werner or George G. Fuchsel, did interesting analyses on the formation of different rock strata, understanding that the layers are ordered chronologically upwards. Georges Cuvier (France, 1769-1832), considered the father of paleontology, introduced the idea that most animal fossils in the different strata were remains of species that were now extinct. He proposed that many of the geological features of the earth and the past history of life could be explained by catastrophic events that had caused the extinction of many species of animals. There was not a single catastrophe but several, resulting in a succession of different fossils. Charles Lyell (UK, 1797-1875) also developed similar ideas on the perpetual change on the features of the earth, eroding and reforming continuously. These discoveries opened up the idea that the earth has a history, a long history full of different events rather than, as per the traditional view, being a static body, created all at once.

In 1862, the British physicist William Thomson (who would later become Lord Kelvin) published calculations that fixed the age of the earth at between 20 and 400 million years (England et al., 2007). He assumed that earth had formed as a completely molten object, and determined the amount of time it would take for the near-surface to cool to its present temperature (WP). Kelvin also calculated, in 1892, the age of the earth by using thermal gradients, and arrived at an estimate of 100 million years old (England et al., 2007). His calculation of the cooling age of the earth was not far from the lifetime of around 20 million years of a solar mass star producing energy with gravitational contraction as its only energy source (Martínez & Trimble, 2009), as calculated by Helmholtz and Simon Newcomb, so this encouraged Kelvin to be more certain of his results, and even to restrict his date to within 20-40 million years, in 1897, with a higher probability of it being closer to twenty. Other scientists backed up Kelvin's rough calculations. George H. Darwin, Charles Darwin's son, proposed that the earth and Moon had broken apart in their early days when they were both molten. He calculated a value of 56 million years as the time necessary to get the current 24-hour period of rotation due to tidal friction. Also, in 1899 and 1900, John Joly calculated the rate at which the oceans should have accumulated salt from erosion processes and determined that oceans were about 80 to

100 million years old (WP). This is one of many examples in science of how an incorrect result is believed to be correct just because there are two or more independent confirmations of it. Indeed, one wrong result invites another wrong result. And two or more independent calculations of a number may give the same result just because of two or more wrong assumptions which get similar numbers by chance, or because of the influence of the previous known results. In the present case, Kelvin's calculations of the cooling age of the earth did not take into account the convection inside the earth, which allows more heat to escape from the interior to warm rocks near the surface; and Kelvin's calculation using thermal gradients did not realize that the earth has a highly viscous fluid mantle; also, the calculation of solar energy did not take into account the as yet undiscovered nuclear fusion reaction, and the other calculations supporting ages lower than 100 millions of years were based on wrong assumptions too.

Fortunately, scientists closer to natural history did not accept those time spans. A time of 100 million years seemed too short to be plausible. In 1869 Thomas H. Huxley attacked William Thomson's calculations, claiming that, although they appeared precise in themselves, they were based on faulty assumptions. In 1895, John Perry produced a model of a convective mantle and thin crust which gave the earth an age of 2-3 billion years. Geologists would realize soon after its discovery in 1896 that radioactivity could provide a method for the determination of the age of the earth, giving ages much greater than 100 million years, and that they also had another heat source to take into account if they wanted to determine the cooling time of the earth.

In radioactive decay, an element breaks down into another, lighter element, releasing alpha, beta or gamma radiation. A particular isotope of a radioactive element decays into another element at a given ratio. Some radioactive materials have long periods of decay and the determination of the relative proportions of radioactive materials in geological samples may be used to measure their age. The chemist Bertram B. Boltwood produced some initial results dating the age of some rock samples in 1907. In the 1910s and 1920s, the radiometric dating method would be established by Arthur Holmes, who was particularly interested in its application to geology, giving the earth ages of 1.6-3.0 billion years. This method was claimed in 1931 to be the only reliable mean of pinning down geological time scales, though it would be improved in later decades.

Today's accepted age of the earth, 4.54 billion years, was determined by Clair C. Patterson (Patterson 1956), using uranium-lead isotope dating on several meteorites, a determination which has not changed in the last fifty years, giving the impression that the race to establish the age of the Earth is now over.

Another interesting episode of geological research concerning the history of the earth is the discovery of plate tectonics. Even in the early twentieth century, geologists assumed that the earth's major features, such as mountain ranges, were fixed, and most of them could be explained by vertical crustal movement. With the arrival of the research scene of radioactivity, once it was known that the earth was billions of years old and its core still sufficiently hot to be liquid, it was possible to elaborate a theory which explained facts such as the apparently similar shapes of the opposite coasts of the Atlantic Ocean, in Africa and South America, for instance (a fact already observed with curiosity by the cartographer Abraham Ortelius in 1596). Plate tectonic theory arose in 1912 out of the hypothesis of continental drift proposed by Alfred Wegener (Germany, 1880-1930). He suggested that the present continents once formed a single land mass that drifted apart, thus releasing the continents from the earth's core and likening them to iceberg of low density granite floating on a sea of denser basalt (WP). This was a bold proposal which was, however, not recognized by the scientific community until the 1950s, after further evidence showed that Wegener was right.

Wegener is perhaps one of the most important heroes of the recent history of science, claiming something very important, relating to ordinary experience nothing less than the movement of continents, and being ignored at the time, as happens with many of the important truths in science. One can imagine the satisfaction of the scientist in elaborating his theory while looking at the shapes of the coast of West Africa and the east coast of South America on a world map. One may guess that the word "Eureka!" was in his mind. Indeed, Wegener developed this idea not just as a simple inspiration from looking at maps (it was in 1910 when he realized the coincidences in a world map) but also by analyzing information from different disciplines: palaeontology, biology, paleoclimatology, geology and geodesy. It was not indeed the lack of proof in the 1910s which caused the theory to be mostly rejected but the lack of a convincing proposal of a theoretical mechanism to explain plate tectonics. Unfortunately, this is something very common in science, too: the scientific community usually is reluctant to believe what the experiments or observations show up until there is a theory which explains the facts. But it is the theory which should be a servant of the empirical data and not vice versa.

2.6   Evolution theory

I have previously described theories which were very important to science: heliocentric astronomy, atomic theory and others. I am a physicist and I am an enthusiast for the revolutions brought about by these theories. In my opinion, however, the most important of all scientific revolutions is the theory of evolution. Why? Because this revolution affects not only our concept of nature, our concept of matter or the position/motion/age of our planet, but also our understanding of ourselves. It is the answer to questions such as: Who are we? Where do we come from? Its importance surpasses the limits of science or pure knowledge. Of course, the debate about religious beliefs was and is still important, but it is more than that. It is perhaps the scientific theory with the greatest relationship to the humanities in general, philosophy, anthropology, history (prehistory), etc. It is a pinnacle in human wisdom, perhaps the highest.

The theory is well known and I will not spend time describing it. There are two major points to it: 1) Living systems or organisms are divided into species; these species are not static but evolve, giving rise over long periods of time to new species, an idea which was already in the mind of many naturalists at the beginning of the nineteenth century; 2) The mechanism of evolution is natural selection, by which individuals with advantageous hereditary characteristics to fit the environmental conditions have a higher probability of surviving and producing descendants, the survival of the fittest. These ideas were developed by Charles Darwin (UK, 1809-1882) and Alfred Russel Wallace (UK, 1823-1913). The explicit mechanism for transmitting hereditary features was not known to these naturalists but would be found later through the discovery of genes and their mutations. The laws of genetic heredity were discovered and published in 1866 by Mendel; another case of a scientist whose ideas were not paid attention to in his own time and which would have to be rediscovered later, in 1900. The microscopic description of the genes embedded in ADN molecules was presented in the 1950s.

Darwin's work is an excellent example of science carried out with patience and care. As a naturalist he compiled information during the five years in his trip around the world, in the ship *HMS Beagle*, starting in 1831. Later, he tried to understand the things he had observed. One spark of inspiration for the idea of natural selection was Darwin's reading in 1838 of a book by the economist, Malthus, *An Essay on the Principle of Population*, which talked about how human populations grow faster than any power on earth to produce subsistence for man, so there is a fight for survival and there is misery, sickness and finally death for those who are not so competitive. Darwin would extrapolate this idea and apply it to all species of animals and plants. This illustrates to us how important it is that a scientist has a wide cultural background and is interested in reading ideas from other fields, because many times inspiration does not come from moving in a closed circle of ideas, within a laboratory or a specialized research centre, as is the case in present-day research. Curiously, at the end of the nineteenth century Darwin would inspire other economists and moralists through a newly revived Malthusian idea applied to human beings and called "social Darwinism". In 1842 he had already practically finished his research on the idea that the origin of species lay in natural selection. However, he would wait for another seventeen years before publishing his work because he wanted to improve some points and gather further proofs.

Wallace was a very different kind of scientist. He also travelled in South America, Oceania and Asia as a naturalist, doing extensive field work. In 1855, during his stay on the island of Borneo, he wrote *On the Law Which has Regulated the Introduction of Species*, in which he expressed the idea that species should be evolving with time, although without specifying the mechanism by which they would do it. Three years later, in 1858, he began to think about Malthus's book, which he had also read, and he arrived at the same conclusion than Darwin, that the changes take place regulated by what he called "survival of the fittest". Quickly, he wrote a paper on the question. Wallace was effectively a different kind of scientist when compared to Darwin, the kind of character that has bright ideas from time to time and wants to express them to the community very quickly, although the idea is not fully developed. As a matter of fact, Wallace was a very prolific author, publishing during his life twenty-two full-length books, 508 scientific papers (191 of them published in the journal *Nature*) and more than 200 other shorter pieces. He worked on biogeography, natural history, evolution theory, anthropology, social questions, and spiritualism and phrenology. Both types of thinker are important for the development of ideas: individuals like

Darwin who spent most of his life developing one of the most important scientific ideas to a high degree of perfection, and restless individuals like Wallace who work in many fields and have many ideas which they have no time to develop and just deliver as seeds others to work on. They are two different types of genius.

At that time, scientists usually sent a paper directly to their colleagues to get an opinion on it before publishing it in a journal. Nowadays, as I will explain in later chapters, journals themselves contact referees (mostly anonymous) to review papers and judge whether they deserve to be published or not. In 1858, Wallace sent his manuscript to Darwin, who was surprised to find that the same ideas he had been developing over many years were in Wallace's paper. Wallace was indeed one of the correspondents whose observations Darwin had used to support his own theories, and the two scientists had exchanged correspondence previously. Darwin was an honest scientist and, although he had being working in the theory for a long time, and he had witnesses to prove it, he had not tried to take all the credit for himself. Wallace was also honest in recognizing Darwin's priority, and he would become a defender of Darwinism. How different these characters are by comparison with the arrogant Newton! Darwin sent the manuscript to other colleagues and publishers with a letter saying "he could not have made a better short abstract! Even his terms now stand as heads of my chapters [...] he does not say he wishes me to publish, but I shall, of course, at once write and offer to send to any journal". Although Wallace had not intended to publish his paper, thanks to the recommendations by Darwin it was published in the *Journal of the Linnaean Society* in 1858 with some excerpts of letters and private communications that Darwin had sent in 1847 and 1857 about his theory, which established his priority. The following year, in 1859, Darwin would publish *On the Origin of Species*.

There were some small differences between the ideas of Darwin and Wallace. Darwin, for instance, emphasized competition between individuals of the same species to survive and reproduce, whereas Wallace emphasized environmental pressure on varieties and species forcing them to become adapted to their local environment. More important divergences between the ideas of the two scientists would come later, in the application of the theory to human beings (WP). Wallace would move towards spiritualism (even geniuses get lost in stupid ideas; Newton also dedicated an important part of his life to alchemy or theology), and he maintained that natural selection cannot account for mathematical, artistic, or musical genius, as well as metaphysical musings, and wit and humour. Wallace eventually said that something in "the unseen universe of Spirit" had interceded at least three times in history. The first was the creation of life from inorganic matter. The second was the introduction of consciousness in the higher animals. And the third was the generation of the higher mental faculties in mankind. He also believed that the *raison d'être* of the universe was the development of the human spirit. These views greatly disturbed Darwin, who argued that spiritual appeals were not necessary and that sexual selection could easily explain apparently non-adaptive mental phenomena. Wallace's views in this area were at odds with two major tenets of the emerging Darwinian philosophy, which were that evolution was not teleological (without finality) and that it was not anthropocentric (WP).

Darwin's success was important not only because for his proposal of the mechanism which produces the evolution of species, but because of its philosophical implications. And, in particular, because of his thoughts about the nature of human beings, an area in which Wallace was lost among superstitions. Science was a dream of reason, initiated by those ancient Greek physicists who tried to remove the mythological elements from the phenomena we observe around us. This is perhaps the highest pinnacle of achievement in this vision. Finally, human beings could be explained in scientific terms, and our characters could be compared with the rest of the universe, as part of a blind nature without purpose. Freudian psychoanalysis reached a comparable peak, but in less scientific terms.

Certainly, there remain even today some detractors of Darwinism (though not usually biologists) who are disturbed by its implicit materialist ideas. In the most cautious cases, religion moved aside and left the question of the origin of humans to naturalists. However, there were, and still are, some religious fanatics who have attacked the theory just because it upsets their beliefs. There are certain subtle technical questions in Darwin's ideas which are subject to debate: that the process of evolution is not as continuous as Darwin thought (Gould, 1989), the different sources of mutation, etc. But the main aspects are quite solid; at least, nothing better can be found in scientific terms. Ideas like "intelligent design" or similar are not science. And using scientific arguments to claim those non-scientific standpoints (Hogan, 2004, ch. 1; Dolsenhe, 2011) seems pathetic.

2.7　Chemistry of life

Along with the theory of evolution and genetics, the other great pieces in the puzzle to understand the nature of life were biochemistry and molecular biology, which made their greatest developments during the twentieth century. Their purpose, to explain life in terms of physics and chemistry, is a great reductionist enterprise.

The idea that life is animated by vital forces which make it different from other phenomena in nature, "vitalism", lasted until the end of the eighteenth century, except perhaps among some materialists and followers of Democritus's theory of atomism which believed that everything is reduced to atoms and their motion. Within the contemporary era, important steps were made in understanding that vital laws are ruled by common physics. For instance, in 1791 Luigi Galvani discovered the relationship between electricity and muscular movements in animals, while in 1847, the physicist, mathematician and physiologist Helmholtz established the principle of the conservation of energy in living beings. Nevertheless, the main reason for discarding vitalism would be the understanding of the substances of which living beings are composed. Still at the beginning of the nineteenth century, organic chemistry was usually thought of in vitalist terms, but the situation would change during the century, with discoveries such as the synthesis of urea from inorganic material by Friedrich Wöhler in 1828, or formic acid by Marcellin Berthelot in 1856. The fundamental conception that underlay all Berthelot's chemical work was that all chemical phenomena depend on the action of physical forces which can be determined and measured (WP). He offered opposition to the vitalist attitude, and by the synthetic production of numerous hydrocarbons, natural fats, sugars and other bodies he proved that organic compounds can be formed by ordinary methods of chemical manipulation and obey the same principles as inorganic substances, thus exhibiting "the creative character in virtue of which chemistry actually realizes the abstract conceptions of its theories and classifications—a prerogative so far possessed neither by the natural nor by the historical sciences" (WP). Knowledge of the structural composition of many organic substances had been developed since 1845 by Justus Liebig and his collaborators. It was also important for the understanding of those substances which constitute living and non-living beings. Among serious biologists, vitalism as a belief was vanishing, and it was understood that organic chemistry could be expressed in the same terms as inorganic chemistry (López Corredoira, 2005, sec. 3.2).

The earliest beginnings of biochemistry would come with the discovery of the first enzyme, amylase, in 1833 by Anselme Payen. There were plenty of interesting discoveries in this discipline which would reveal the chemistry of life. One of them was the proposal in 1902 by Emil Fischer and Franz Hofmeister that proteins were composed of chains of amino acids. I am also amazed by the synthesis of amino acids in simulations of the conditions of the primitive atmosphere of the earth in the 1950s (Calvin, 1956; Fox, 1956). Another relevant historic event related to biochemistry and molecular biology is the discovery of the gene and its role in the transfer of information contained in cells. In the 1950s, James D. Watson, Francis Crick, Rosalind Franklin and Maurice Wilkins were instrumental in establishing the double-helix structure of DNA and suggesting its relationship with the transfer of genetic information. This was perhaps one of the last important pieces of the puzzle of the chemistry of life.

The history is long and contains plenty of heroes. Unlike other fields of science, I do not think we can speak of only one name being associated with the genetic revolution, not like Darwin's description of the theory of evolution, for instance, which marked an abrupt transition in the way that biology was carried out. In this instance, there was a gradual development, a collective creation which culminated in a magnificent achievement: knowledge about the structure of life. These branches of biology (biochemistry, molecular biology, genetics) are also perhaps the sciences which have made the most important developments in the second half of the twentieth century. I think this is probably due to their late beginning, as in the arts. Think, for instance, about cinema in comparison to the other arts; the later they begin to develop, the later they begin to decline. Anyway, the golden age came late for these branches of the sciences, but has already passed away. It is difficult to imagine that these sciences may in the future offer something as revolutionary as an accurate understanding of the philosophical concept of life, reduced to physics and chemistry. There remain many small details to be made explicit, but the concept will not change. When sciences reach their limits as producers of new ideas they are claimed to be in a successful golden age, thanks to the application of technology. That is the case with the Genome Project and other megaexpensive projects, whose new discoveries are fed by technological applications rather than new ideas.

2.8 Medicine

The applications of science are indeed one of the most important causes for their existence. For less intellectual people (i.e., most of society), who don't appreciate the merits of science as knowledge for the sake of knowledge, the development of applications to make our lives more comfortable or to provide plenty of new possibilities for market consumption is much more important. This is called progress, and it is usually associated with science. In my opinion, however, technological progress, particularly since the industrial revolution in the middle of the nineteenth century, is something negative. Machines have made society uglier in aesthetic terms, they have created serious ecological damage to the environment, they have increased the possibility of destruction during wars, etc. They force civilizations to work against the nature of human beings. Technology in coalition with the dark forces of capitalism is becoming a monster, difficult to control, threatening to devour all humans.

There is one important technical application which deserves greater consideration: medicine. Sickness and death are part of human existence, but human beings can go beyond nature, to favour their existence as individuals rather than simply being part of an ecosystem. As a consequence of the development of medicine, among other things, we now have an overpopulated world; not very positive, either: Again, we are forcing humans to go against their nature, breaking the rules of natural selection and the equilibrium ecosystem.[9] But who wants to live without medicine? Apart from members of particular religious sects, who wants to die from a curable illness just to avoid disturbance in the ecosystem? Answering these questions gives us the reasons for the development of this science, and the meaning of its existence.

---

[9] Indeed, nothing is against or apart from nature because we, human beings, are also part of nature, but a very special part, with some peculiar rules of its own, separate from the rules which dominate the rest of the species. Even what is called "cultural selection" is not escaping natural selection, but adding some new twists to the meaning of "survival of the fittest". No new natural laws emerge with regard to human beings; they are the same rules as for the rest of nature but with a higher level of complexity.

I have already talked about medicine with reference to Hippocrates in subsection 2.1. Equally important in ancient culture is Galen. There are indeed plenty of important names in the history of medicine right up to the present day, and it is impossible to include all of them in this subsection. I will speak only about five significant people, five great benefactors of humanity, who contributed to saving it from terrible illnesses: Jenner, Pasteur, Ehrlich, von Behring, and Fleming.

Edward Jenner (UK, 1749-1823) brought to medicine the method of vaccination, which would serve to prevent many illnesses, some of them fatal. In particular, he used vaccination to prevent smallpox. At the end of the eighteenth century, many people contracted smallpox, which was fatal in 10% of cases, and left pockmarks in the rest of the cases, usually disfiguring the faces of those affected. There was a superstition among farmers that once a person contracted the cowpox, he or she would not contract smallpox. Cowpox was transmitted from animals, usually cows, hence its name, to humans, and was quite benign, just producing some pustules in the hands, while smallpox was passed from human to human. Jenner decided to investigate whether the farmers' belief was real or not, observing the transmission of the infection, and even doing some experiments with humans. In the end, he concluded that effectively once a person has had cowpox, he or she becomes immune to smallpox. Thus, the idea came to his mind of infecting deliberately people with cowpox; this could be done using liquid taken from the pustules of someone infected with cowpox: that is, vaccination.[10] Today we know why this method works in terms of the immune system of our bodies. At the time when Jenner delivered his results and solutions, in 1798, it was just a phenomenological observation, but obtained with care and through systematic analyses. In fact, Jenner was not the first one to discover this solution: As noted, farmers already suspected that becoming infected by cowpox was a guarantee of avoiding smallpox, but it was also usual in Turkey, at the beginning of the eighteenth century, for people to be inoculated against smallpox with liquid taken from the blisters of people with mild symptoms of smallpox. Possibly, the discovery of vaccination should be attributed to popular wisdom. Nonetheless, credit must go to Jenner for carrying out an extensive analysis with rigour rather than just listening to popular rumours. This is indeed what distin-

---

[10] "Vaccination" comes from the Latin name for cowpox, "vaccinia".

guishes western medicine from other healing practices. Quacks and witch doctors may have efficient methods available to them, and western medicine may have indeed learnt something from them, but the success of the latter stems from scientific methods which are able to test and separate the efficient solutions from mere superstitions, and provide a naturalistic explanation for them rather than mythological, religious or superstitious ones.

Louis Pasteur (France, 1822-1895) was perhaps most responsible for the theory that infectious diseases are transmitted by living microorganisms, or germs, which can move from one person to another. This idea opened up a wide field of new possibilities in searching for treatments for many of those diseases investigated by Pasteur and Robert Koch. Again, as with atomic theory, we have the discovery of invisible elements whose reality is nonetheless indubitable. And this reality is of such extraordinary importance that it is nowadays difficult to understand how physicians could have thought about illness without this knowledge. Indeed, a proper understanding of what an illness is starts here, with modern medicine after the middle of the nineteenth century. The contribution of Pasteur was not only theoretical but also methodological. Among his numerous contributions to the fight against germs is "pasteurization", which consists of the heating of foods to kill the germs, in order to facilitate a longer period of preservation. Pasteur demonstrated definitively that putrefaction was caused by the presence of living microorganisms and that it would be stopped if germs were completely eliminated and the substance isolated from new sources of contamination, as opposed to the prevailing belief in the spontaneous generation of living microorganisms from inert matter. Some people thought at the beginning that Pasteur was wrong because meat, even when boiled, could become rotten; Pasteur argued that this was due to the presence of germs in the air which contaminated the meat after pasteurized. If the meat was kept in air free of germs it would not become rotten, as he indeed proved in experiments. In the 1870s, he also persuaded physicians to boil their instruments and dressings before applying them to the wounds of patients. The practice of disinfection had also been initiated before Pasteur, by other physicians, such as Ignaz Semmelweis in the 1840s, who recommended that physicians to wash their hands with calcium hypochlorite before touching the patients, something which was regarded as absurd by their colleagues at that time (Gillies, 2008, ch. 2), or the surgeon Joseph Lister who, in the 1860s, applied carbolic acid to wounds or surgical incisions and obtained a very significant reduction in postoperative deaths.

Paul Ehrlich (Germany, 1854-1915) and Emil von Behring (Germany, 1854-1917) were the founders of serum therapy and discoverers of antibodies in the immune system. Behring discovered that animals produce chemical substances which destroy germs, and that these antigens were created the first time an animal contracted an illness. In 1889, Ehrlich and Behring worked together, and it was Ehrlich who discovered how to encourage animals to produce antibodies, by deliberately infecting them with diseases; after that, one could extract some blood from the animal, concentrate the antibodies in a serum, and inject this into human beings, allowing them to be immunized without becoming sick. In 1892, the two men found a diphtheria antitoxin, and together with Émile Roux obtained a serum therapy against diphtheria. Behring would also develop a serum therapy against tetanus. Ehrich would develop his side-chain theory in 1897, explaining the effect of serum, and would then experiment in a new area he would create: chemotherapy. He tried to find a chemical substance which killed germs but which was inoffensive to animals, making many experiments with an army of collaborators to test hundreds of substances containing arsenic, which in principle acted as an effective destroyer of trypanosomes in some germs. He found that atoxyl, an arsenic compound, was not toxic to trypanosomes but it proved useful in treating sleeping sickness. In 1909, he and his student, Sahachito Hata, developed "Salvarsan", a treatment for syphilis, which was much more effective than the usual therapy of mercury salts used until that time.

In 1928 Alexander Fleming (UK, 1881-1955) was investigating the properties of staphylococcus, a genus of Gram-positive bacteria. Before taking his summer holidays, he left the samples of staphylococci in a corner of his untidy laboratory. When he came back, he noticed that one of the samples had become contaminated with a fungus, and that the colonies of staphylococci that had immediately surrounded it had been destroyed, whereas other colonies further away were normal (WP). Fleming identified the mould that had contaminated his plates as being from the Penicillium genus (Diggins, 2003). This accidental discovery and the subsequent isolation of penicillin marks the beginning of modern antibiotic treatments. He later investigated its positive anti-bacterial effects on many organisms, and noticed that it affected many other Gram-positive pathogens, which caused scarlet fever, pneumonia, meningitis, diphtheria, and the Gram-negative bacteria of gonorrhoea. Many other conditions would also prove to be treatable, thanks to penicillin. In the following two decades, methods to refine usable penicillin would be developed, and a huge number of lives would be saved, thanks to this accidental discovery. This is one among the many examples of fortuitous discoveries, which shows us that the methods of science are not always contained by a straightforward framework. We have to praise Fleming for keeping his laboratory untidy and going off to enjoy his holidays rather than remaining restless in a clean and ordered laboratory, like many workaholics in our own society. Science is the product of work, like many other products of our culture,[11] and the fruit of genius/intelligence, but it is also the result of chance. Sometimes, the observation and interpretation of accidental events around us is more important than the laborious work of systematic experiments.

There are many other important discoveries, not only about the healing of our bodies but about our knowledge of our bodies, for instance in neurosciences and other soft sciences related to psychology. I will skip the vast episode of the knowledge of our brain.[12] As

---

[11] The Spanish cello player of the first half of the twentieth century, Pablo Casals used to say: "Art is a result of work".

[12] Nonetheless, I would like to note that the major discovery, that the brain is the origin of all our thoughts, was already made by Hippocrates twenty-five centuries ago, and the modern discoveries about how the brain works are merely anatomical descriptions of humanistic approaches developed before neurosciences came into being, for instance by Freud (Lakatos & Navarro, 2006). The most fascinating topics in neurosciences, the philosophical topics related to the origin of the "ego", are still a matter of speculation, and it is not probable that neurosciences can give a more complete picture of that, beyond anatomical descriptions.

In a manifesto published in 2004 by the German journal *Gehirn und Geist* (Brain and Mind) (Elger *et al.*, 2004), signed by eleven important researchers in neuroscience, it is suggested that there are three levels in the study of the brain: the upper level, which explains the function of the great areas of the brain, about which many things are already known; the lower level, concerning the processes of neurons and other cells and molecules in the nervous system, about which many things are also known; and the intermediate level which may explain what happens among associations of thousands of cells, and about which relatively little is currently known. According to the manifesto, the great philosophical questions about the brain, such as the emergence of consciousness, will, etc. may be related to these intermediate levels, about which neuroscientist expect to have some knowledge within

mentioned several times, this chapter does not try to be comprehensive in listing scientific highlights. Thanks to all the discoveries in medicine a large percentage of deaths could be avoided, and infant mortality much reduced. This would produce, as a consequence, an increase by a factor of four in the world population in only one century, in spite of two world wars. This is leading us to the problem of overpopulation, although this problem is also being solved with one of the most revolutionary discoveries for the civilized world: contraception. Let's see whether the future development of this intellectual game is able to create an equilibrium between human life and death.

2.9   Socioeconomic conditions of scientific advances

These examples of scientific highlights throughout this chapter are more than enough to recognize the important role of natural sciences for our world. Apart from the technological applications, science is also a very important part of our culture, of the soul of our western civilization. There is a sense to its development; there are indeed many senses. The nonsense would be its non-development when there is the power of intelligence to do it. Scientific research has been conditioned by the economic and technical necessities of each epoch (Hessen, 1931), it was not the product of the pursuit of pure knowledge in a society, but its results went far beyond mundane needs, becoming one of the most important parts of human culture.

There is no economic value associated with the worth of science. It has no price or market value, and if it did have one, it would be an astronomical number, beyond the amount of money presently available on the planet. Nonetheless, one may wonder about the amount of resources necessary to develop science during its history. An accurate calculation of it would be complex and would need lengthy research. However, roughly speaking, we can say something general. The resources dedicated to research in pure science (neither applied science nor teaching) until a century ago are totally negligible

---

twenty to thirty years (counting from 2004). It is a good argument for getting further money from the state, to fund neuroscience. However, in my opinion, the important philosophical questions about the brain will not be solved even within a hundred years. All neuroscience can provide is more anatomical descriptions, while the philosophy of our brain will be the same one that materialists already discussed many centuries ago.

in comparison with the resources devoted to it in our own society now. The investment in science from states and private industries would be more significant after the industrial revolution but mostly for technological applications rather than foundational science. The investment in research in foundational science would become higher with the creation during the twentieth century of many research institutes, or the promotion of research in the departments of many universities. There are also some precedents for the development of science in various leading countries (U.K., France, Germany and a few others) in the nineteenth century but very few could be considered to have made a major investment in resources. At present, this investment is not negligible at all: it takes up to 3% of Gross Domestic Product (GDP) in some of the leading countries in scientific research, and at least some percentage of GDP in most countries of the world, which is not negligible at all: we are talking about a trillion dollars per year throughout the whole world (OECD data). Of course, much of this research includes everything under the name of Research and Development (R&D): applied science and military projects, but a good ratio of it is for pure sciences too. Around 20% of R&D funds are dedicated to research in pure sciences, so we are talking about 200 billion dollars per year.

Economically speaking, we can say that roughly 99% of the research efforts have been made in the last few decades, while the remaining 1% is accounted for by the rest of the history of humanity (taking into account corrections for inflation and economic parameters to make the economies of different currencies at different times comparable). However, almost all of the most important highlights of pure sciences were achieved with that 1% of investment. It is a datum to take into account in our reflection about the sense of promoting fundamental research: exponentially increasing expenses with exponentially diminishing returns. I will come back about this point in future chapters.

We may wonder why science in the past was so cheap, or why it is now so expensive. There are three reasons:

1) The ratio of scientists to general population in the past was much lower than now;

2) The instruments needed by empirical scientists in the past were much more rudimentary and easy to build, and consequently cheaper;

3) Most of the scientists did not receive a salary for their research. In most cases they even paid their own research expenses,

and they did not have as many paid vacations disguised as "conferences" as they do now.

The number of scientists and engineers dedicated to R&D in the present day is 5.8 million (OECD data, 2006, excluding India); that is, on average, around a thousandth part of the world population. In developed countries the ratio is much higher: for instance, in U.S. it is around one in two hundred. Let us assume that the number of people dedicated to research in the pure sciences is again 20% (in the same ratio as the funds): we will have an average ratio of $\sim 2 \times 10^{-4}$ people dedicated to research in pure science. In the whole of the nineteenth century, the number of persons around the world who published at least one paper in a scientific journal, either pure or applied science, was approximately 115,000 (Gascoigne, 1992), although most of them only published one paper and did not dedicate their lives to science. The number of people who lived through the hundred years of the nineteenth century was around 3 billion, so a ratio of scientists to general population of $\sim 4 \times 10^{-5}$ stems from that census, even including people who only produced one paper, is lower than the current one. The number of scientists in the seventeenth century is roughly ten times lower than in the nineteenth century (from the analysis by Gascoigne, 1992, of scientists who appear in the *Historical Catalogue of Scientists and Scientific Books*) while the population was roughly half that of the nineteenth century, so the ratio in the seventeenth century would be circa 10^{-5}.

That expenses for instruments are much higher now than in the past is quite obvious, and I will not give further explanations here. Certainly, the telescope which Galileo built was much cheaper than the 42m E-ELT telescope to be built by European Southern Observatory in Chile with a planned cost of one billion Euros, and space telescopes are even more expensive. I will discuss this further in future chapters.

The third point is more interesting to comment on. Certainly, the funding of science has changed a lot nowadays with respect to the past. In the next chapter, I will talk about the present situation, but now I will illustrate, drawing on examples of particular scientific advances, how research was previously funded. In general, the answer is that there was no specific channel for funding science. The motor which drove science was mainly the passion for science itself in a few individuals. Kings, aristocrats or rich people might provide resources to develop science, but this was very irregular and not subject to systematic predetermined programs. Indeed, the most important advances in science came without any previous economic support. Of course, scientists had to earn money because they had to eat, they needed a house, etc. but their income came from other sources. Some scientists belonged to the aristocracy and did not have to work for a living. Others taught in the university; only in the late nineteenth century would it become usual for universities to support research and the creation of laboratories. And a few were supported, as noted, by wealthy patrons. The professionalization of research is actually a very recent thing.

Each biography of a scientist needs to be considered differently, and I shall not carry out a full analysis of the socioeconomic conditions of science throughout history here. In general, the very best scientific advances were obtained under very modest economic conditions, with a total absence of external support, or else were self-funded, although there were also scientific projects with high costs, for instance, the organization and management of the vast Harvard Observatory surveys of the apparent magnitudes and spectra of a very large number of stars carried out in the late nineteenth century by Edward Pickering (Longair, 2001). During the years when the great surveys were undertaken, Pickering obtained several hundred thousand dollars from the Henry Draper Fund, $400,000 from the Paine Fund in 1886, $230,000 from the Boyden Fund in 1887, and $50,000 from the Bruce Fund. It is estimated that Pickering's personal contribution to the project amounted to about $100,000.

A remarkable example of self-funding is Darwin, who never had a remunerated academic position. He had a private income and was comfortably off. His five-year trip on board *HMS Beagle* as naturalist was also self-funded. His latter years would also be dedicated to research, conducted at his house in the country, and, as noted, he prompted perhaps the most important of all the scientific revolutions.

Another example may be observed in the biography of Santiago Ramón y Cajal (Spain, 1852-1934), who discovered how the nervous system in vertebrate animals operates, by identifying neurons as the nervous cells which transmit information, and disentangling the mechanism which is nowadays the basis of neuroscience. His first steps in research were carried out with a microscope he bought with his own money in 1877; eight years later he would receive a better microscope from a public governmental organization as a reward for his work on cholera, and with it he would make his greatest discoveries, which would lead him to receive the Nobel Prize in Medicine in

1906. Cajal had survived on the modest salary paid to professors of Spanish universities at that time. Significantly enough, in 1900, once he had become a celebrity, receiving many prizes recognizing his achievements, obtained mostly in a private laboratory at his house, he was named director of a fully equipped research institute funded by the Spanish government. One of the first thing he did was to reduce his salary, because he did not want to become an affluent person and thought that it was unfair he should receive a high salary while people in his country were working hard to pay the taxes used to fund the research institute. What an example of humility from an exceptional scientist! At present, the more mediocre scientists are, the more money they ask for; even with huge budgets, they never have enough. Even when there is an economic crisis and the country has got problems, rather than being more modest in asking money, scientists claim even more money because they think research is a fundamental thing for the development of a country, and to escape the crisis. Apparently, science has become an egoist system which thinks only of retaining its status, even when this is against the interests of society.

Among the recent highlights of physics, in the twentieth century, we have the very famous case of Einstein. Einstein graduated in 1900 with a degree in physics. He tried to get a position as assistant to Professor Heinrich Friedrich Weber at the university in Zurich , but he failed to get it, both because Einstein's grades were not very good and because Weber was offended to be addressed by Einstein as "Herr Weber" instead of "Herr Professor Weber", a serious failing in a German-speaking university at that time (Miller, 2001; Gillies, 2008, ch. 9). Instead Weber appointed two mechanical engineers. Einstein later enrolled to do a Ph.D. thesis with Weber, but things did not go any better and in 1901 he abandoned it it. He tried to get a position as assistant with other professors or to submit a Ph.D. thesis to the other university in Zurich but this also failed, so in 1902 he began work at the Swiss Federal Patent Office in Bern. During free time while working in the patent office, Einstein would develop his four famous papers, published in 1905 including "On the Electrodynamics of Moving Bodies", which outlined Einstein's theory of special relativity. He would earn a Ph.D. with one of his less revolutionary papers. Later, he would be recognized for his works and would be given a position as researcher, producing the very important theory of general relativity from a paid position as professional researcher. At the pinnacle of his glory, Einstein spent the last twenty years of his life, in a very good position at the Institute of Advanced Study, at Princeton, New Jersey, but he produced nothing of importance.

The other great theory in twentieth-century physics, quantum mechanics, deserves a similar consideration. As was said by Gregory Chaitin in an interview John Horgan (1996, ch. 9): 'Remember that quantum mechanics, which is such a masterpiece, was done by people as hobby in the 1920s, when there was not funding'. In fact, the people who developed quantum mechanics were professional physicists, there were not outsiders, and they worked mostly in academic positions, but there was no specific funding for a line of research whose purpose was to explain observations in the microscopic world. This shows us how the good ideas emerge when they have to emerge, not when there is a financial investment waiting for them. Science products are not like industrial products. And they do not wait for the approval of funds after a bureaucratic process. Of course, experimental and observational science becomes better with better instruments, but instruments alone do not think. As we will see in the next chapter, things nowadays are very different. It would be a shame for formal science if a young man working in a patent office developed a better science than that produced in official places, and the scientific community is nowadays doing its best to ensure this does not happen again, by excluding outsiders.

3 THE INSTITUTIONALIZATION OF SCIENCE AND ITS NEW SOCIOECONOMIC CONDITIONS

During the twentieth century, many countries would join in the enterprise of institutionalizing science. This is customary with the best ideas. It is always the same story: first, a few individuals develop the basics and the most important ideas of a discipline, working against the currents of their times. Second, there is a stage when society begins to admire the intellectual achievements of those individuals and/or realize the benefits of them for our civilization. Third, society supports these activities by creating specific channels of funding for the development of those ideas, and creating a bureaucratic structure around them. One very obvious example is to be found in religion: in the Christian religion, there was a first stage in which Jesus and his followers invented a new revolutionary way of seeing the world; there was then a second stage in which many people could feel the spiritual benefits of that belief, and then came the institutionalization of the church, with its hierarchy of powers. Christ was a poor fellow with great ideals. The highest representatives of the church have been individuals with great power and wealth, living at the expense of the creativity of that poor fellow. The problem with this third stage is that individuals in the organization begin to worry less and less about the important ideas, and more and more about how to get a higher position within the organization. The high ideals of truth or beauty or goodness are very often perverted into a pursuit of personal interests within the organization, and this is usually justified as a necessity for the survival of the institution and its associated groups. The survival of the institution has taken priority over the survival of the ideals.

Science is now in that third stage. It has gained recognition from society, and it is nowadays one of the centres of power which pulls the strings of our society. It is a new church. Many times, philosophers have compared science with religion. I do not think that comparison is appropriate. Science is an activity very different from religion, and its concepts have an empirical basis which is far beyond the beliefs of religion. Nevertheless, from a sociological point of view, just looking at the social organization, there are certainly some similarities.

My experience as a researcher is not universal. As said, I was embedded mostly in the world of astrophysics. Nowadays, all scientists are specialists and it is not possible to have direct experience as a researcher in all fields. Nonetheless, I have observed some patterns of behaviour in my field which I think are universal in the world of research; indeed, I have contacted researchers in other fields, and they have confirmed that my views are applicable to most of research organizations nowadays. Generalizations are dangerous, and one should always bear in mind that there are exceptions and peculiarities depending on a particular field, institution, country, etc. In this chapter, I will offer my own impressions, to illustrate some aspects of the hierarchy associated with the institutionalization of science.

3.1   Students

The first contact with research takes place when a graduate student prepares a Ph.D. thesis. Here, as in many other disciplines, the system adopts a clear position: "things are as we say; either you take it or you leave it". If one wants to work in research, one must be in the service of a program that is predetermined by the authorities responsible for the system. If the student wants financial as well as departmental support, then his or her role is to be obedient to, and assimilate, the traditions of the department.

In an ironic and cynical sense, and in a way that is not so very far from truth, students are usually referred to as "slaves". The slaves are in charge of doing the most monotonous research tasks in the service of the team for whom they work. In the hierarchy of the system, there are also some other lowly figures: the temporary scholarship holders; they are normally students who have not finished their degrees and, therefore, they are below Ph.D. students. They are usually called "summer slaves" because their contracts are in force during summer months, and there is not enough time for them to learn much about research. Therefore, they are used as cheap manpower: they are given a few days to learn a mechanical task and the rest of the summer to routinely carry it out.

This evaluation of the treatment of students is really not always true. In my case, for instance, it was not. However, I feel certain that, in reality, such exploitation is quite extensive and rather more common than might be desirable. Of course, I am describing what other researchers have told me rather than presenting statistical data published by any official organization. Nevertheless, I judge that the sources are sufficiently representative.

In some cases, the students that do most of the work are not able to write a paper containing their results; the bosses do it instead, as first authors of the paper. Students are told that they do not know how to write their own work. In other cases, when the supervisor sees that things will not have the outcome he or she wants, the supervisor will abandon the student. In some cases, the supervisor steals the student's ideas. In other cases, a student's time as a Ph.D. student is over before the thesis is finished because he or she was exploited by having to carry out other work aside from their thesis, or the boss had no time to attend to the explanations produced by the student. In such cases, the student must struggle to survive while writing up.

Few bosses sit down and work with students. Normally, they spend some time during the early days explaining how to do things. After that, the student must carry out the routine tasks. The boss just provides the ideas, if they have them; otherwise, they just make minor corrections. The student spends weeks or months with the monotonous tasks in the laboratory, observatory, or in front of a computer with annoying calculations or simulations that consume a lot of time, or with analytical calculations, or carrying out biblio-graphical research or whatever. Students closer to an empirical branch of the sciences spend a long time with the instruments. The boss is usually present as well, but only to initially explain to the student how the machine works, or when something unusual hap-pens, or there are extraordinary observations unrelated to the usual routine. Meanwhile the boss manages and generates ideas. The boss will say: "All right, but you could try this, or that, or this other thing". The student will spend a week doing "this". He will spend two weeks doing "that", and he will finally realize that it is not feasible. "This other thing" is surely stupidity, but the boss cannot be convinced of that until it is checked with some analyzes (of course carried out by the student). Finally, they deliver the results to the boss, and the boss says: "these are nice but I prefer things as they were before".

What is wrong with this situation? Isn't it normal that the master teaches the student, and the students carry out what they are asked to do while learning? Certainly, it should be so. The fact is that the new graduate student does not know much about the specific area of work, and must be brought up to date. Nevertheless, they are not novices without any knowledge at all. Usually, they have more general knowledge about their branch as a whole than the specialist, who knows a lot but only about their own specific area. Moreover, students have some advantages over the master in this case: they are more creative, more open—with less prejudice—and can give new and fresh points of view about developing research, instead of following anachronistic traditions that are embedded in the interests of the person who has spent a whole life with one idea. Ph.D. students can produce ideas if allowed to produce them, even off the track that was chosen for them. In this regard, it would make more sense if the monotonous work was in the hands of those whose creativity is exhausted, those aged, reputable experts who will produce nothing but copies of what they have always produced. However, the present-day world of science only understands power: the captain gives the orders to the sailor. Because of this, it is usual for routine work to be carried out by students. This is not so that they may learn (since one learns the first time, but not by doing the same task a hundred times), but in order that they produce. In a few cases, Ph.D. students do their own research alongside their work with their supervisor, and in fields other than the subject of their thesis (I would recommend this to future students), but that is not the most common way.

I was once told an anecdote with regard to this. I do not know whether it was true, but it seems that it is a real case: a student speaks to his supervisor and says "I had an idea". Then the supervisor replies: "Ah! You have time to think?". Is this the way to form future scientists? Having them spend time in a thousand routine tasks without time to think freely? Thinking about how to corroborate, yet again, an idea that originated from a specialist of repute is encouraged. However, spending time thinking about one's own ideas, without permission, is something that is really not encouraged by the system; quite the opposite. Initiative is discouraged with arguments such as what is established is well established. Workers for science are created instead of thinkers.

There are also many cases which show the opposite thing: students who are inept, something more frequent in recent decades because of the terrible systems of education which every year produce poor graduate students. These are cases of students who are only interested in getting a Ph.D., without any real interest in science, and who are unable to produce any scientific results without the help of their supervisors. There are many cases like this, but, in my opinion, these students should be eliminated from the system rather than turning them into cogs in the scientific machinery.

What is wrong with this situation, I continue to ask. Mainly, that industrial (mass produced) science, rather than creativity being encouraged, and the period of optimum creativity of a scientist is exhausted with these ups and downs. We must take into account that, in the long history of science, the majority of great ideas were produced by young scientists. If young students, who could potentially produce new ideas, are used as slaves (or perhaps it is better to say "science workers"), then perpetuation of old ideas and intellectual stagnation are rife.

3.2 Postdocs, permanent positions

After getting a Ph.D., researchers achieve a higher status in the hierarchy of science, known as a "postdoctoral position" or "postdoc". And after that, after that long period of learning and adapting, if the postdoc passes through all the filters of the system, they can get a permanent position, something more or less equivalent to the role of a priest in the church. Of course, even a priest can earn a higher status by becoming a bishop or archbishop, or even pope, and in science similar things happen. Senior researchers, do not all have the same weight or authority in the hierarchy. There exist, as in any university department, certain ranks: chair, department director, etc. Apart from these nominal ranks, there also exist certain power status indicators that are associated with other factors.

The researcher who wants to live on his work in the research world must aspire to a permanent position, that is to say, to become part of the body of state functionaries, well known by all because of their efficiency at all levels. Joking apart, the fact is that the life and motivations of the postdoctoral researcher are oriented towards ultimate purpose of obtaining a permanent position. In the best case, the functionary will still be motivated after the goal is obtained, but in most other cases the opposite thing happens. A student who has just completed their thesis rarely obtains a permanent position immediately, but must first go through several institutions (some of them of necessity in a foreign country, although there are exceptions) with temporary contracts as a "postdoc". I really think that this is one of the cleverest things the system offers, because at least the researcher has some years to look for their own pathways in research and, at the same time, avoid the stagnation that is usually produced by early permanent positions.

"Postdoc" status constitutes a hierarchical position for science workers above the Ph.D. student, but below the researcher with a permanent position. A large part of the routine scientific tasks falls to these people but to a lesser extent than the Ph.D. students since, in some cases, they have their own mobility. In many other cases, however, they are contracted workers in a predetermined program. In the last two decades, I have observed, at least in my speciality, how the number of offers of postdoctoral positions with a free choice of research topic has been much reduced, substituted by positions working on a major project under the orders of senior "priests". This is, in my opinion, a huge obstacle for the creativity of young scientists, an unfortunate trend in the present-day bureaucratized system. Moreover, among people who are going to develop a topic freely, there is also a strong bias against all the applications which propose topics which are not suitable according to the mainstream of normal science. Certainly, this is the perfect way to uphold the power of tradition and to castrate new ideas. No revolution is possible within this system; only an outsider can do it.

I think that travelling to a foreign country is of little relevance. Perhaps it is important in order to become more fluent in English or another language. Nowadays research is quite global by nature. Few things can be learnt in a foreign country that cannot be learnt in one's own. The same things are said and done everywhere, with only minor differences. The type of research to which a student is exposed perhaps has a larger influence than does travelling to a foreign country. Generally, after completing a thesis, the researcher remains in the same type of environment, so they do not learn much that is new.

It must be said that not all Ph.Ds continue as researchers. In many cases, private life obstructs the requirement for mobility in the job. Also, the high level of competition means that only a few can continue. In order to gain a postdoc position, a good CV is required, which is not necessarily associated with genius but with the capacity to work, and the support or recommendation of somebody within the system. Without the appropriate recommendation, a career may be cut short. Hence, it is necessary to look for compatibility in the research world. A way to achieve compatibility is by following the general trends in research without trying to create a critical front. With regard to the CV, it greater weight normally comes from the number of publications in professional journals and their citations. I say number (quantity) rather than quality because this is really the predominant parameter. Quality is valued when the committee evaluating the person who is applying for the position has a specialist in the same field, holding the same ideas (that is, not a competition to defend alternative theories). Since each specialist thinks that their field is the most relevant, the CV will always find support with respect to quality when it is oriented towards the interests of the judging tribunal. Otherwise, it will be just a number of papers and their citations, evaluated by weight, like school tasks.

This way of evaluating the efforts of a researcher will be a constant throughout their working life; either to secure a position or to find money for a project, etc. With regard to postdoc positions, we must see in this evaluation system an indirect pressure on the putative free choice in research to focus on subjects already being worked on. The number of publications with little critical content, as well as short-term compatibility, is an indication of this pressure. These will be the factors used to secure other postdoc positions, or a permanent position, if the person does not leave the research world, or chooses to earn their living in another way. From my own experiences and those of others, I have observed that doors are opened and offers made to those who are servile and uncritical. A lot of work must be produced, but without any great aspiration towards saying something important. To obtain an academic position, to obtain tenure, to be successful in obtaining research funds, etc. it is necessary to conform.

3.3   Publications, referees

The fruits of scientific work are gathered in specialized journals, journals that will be read by other specialists and distributed to the libraries of the research institutes all over the world, costing exorbitant amounts either to publish or to subscribe to, meaning they can only be afforded by the wealthy institutions. It should be mentioned here that the scientific journal business is not a trifling question. Nowadays, the journals are a powerful communication tool, written in the international language of English, and with enviable accessibility. It must be recognized that the present-day system of scientific publication is far superior to that of other academic fields, where the unification of language is not a given, and publications are spread over many smaller journals which are difficult to obtain.

As is well known, control of communications and practice of power are closely related. I do not think that I have revealed anything new with such a statement. Thus the system, far from allowing free publication of results among professionals, works hand in hand with censorship. Theoretically, this control is presented as a quality filter but its functions are frequently extended to the control of power. Those researchers who want to publish in these journals are subject to the dictates of the chosen referee and the journal editors, who will say whether the paper is accepted or not: this is the peer review system.

The system of peer refereeing is relatively recent in science. In continental Europe it was mostly unknown before World War II. Rather, decisions about publishing were taken by editors, based on trusting the unquestionable authority of the author of the paper, or, in the case of the author being some young scientist without reputation, on the recommendations of some senior scientist whom the author directly had contacted. The United States was one of the first countries to adopt the policy of "peer reviewing" and, since science was "Americanized" after World War II, this, together with the use of English as the scientific language, extended to all scientific journals in Western Europe. There is a story that when Einstein first submitted a paper to an American journal, *Physical Review*, in 1936, the paper was rejected after a peer review. Einstein replied to the editor, "Dear Sir, We had sent you our manuscript for publication and had not authorized you to show it to specialists before it is printed" (Kennefick, 2009). Certainly, Einstein was not familiar with the peer review system.

The referee, by choice, is usually anonymous. There are cases in which the editor is anonymous as well, and the only name known is that of the secretary. This fact indicates that the activity is not always honest. If it were honest, the referee would not hide himself behind anonymity or, perhaps, be afraid to be pointed out as the critic. If a referee thinks he/she is giving good advice then his/her name should not be hidden.

Referees can offer their knowledge to improve the quality of the paper to be published, or detect errors in a calculation or analysis, if any, or detect contradictions in data, etc. In principle, the idea is good, and it would be better if the refereeing process were always objective and impartial. I think that this is often not the case. There are many instances in which the fate of a paper is dictated by a conflict of interests rather than the merits of the paper.

From my experience publishing scientific papers in refereed international journals (and I have also been a referee many times), I have observed that the reports of the referees rarely detect errors in calculations or data reduction procedures, because the referees are not patient enough to carry out the calculations again or check the codes. This was also observed in an objective experiment: peer reviewing was tested for its ability to detect errors in scientific papers about neuroscience. A paper with eight deliberate errors was sent to 420 reviewers. The average number of detected errors was two, nobody detected five or more errors, and 16% of the referees detected no error (Rothwell & Martyn, 2000).

Apart from minor details—changing the structure in order to better present the argument, rewriting a paragraph to be clearer, citing some other paper (in many cases the referee advises the author to cite some paper of their own, or by collaborators, etc.)—objections are very often to do with the referee's own opinion or how convinced they may be about the content of what is going to be published. Generally, according to my experience and that of people I have spoken to, the more controversial the topic, and the more of a challenge it is to established ideas, and the newer the approach, then the more difficult will be the problems in publishing it, and the higher the probability of its being rejected. Gillies (2008, ch. 2) argues that when a researcher makes an advance which is later seen as a key innovation and a major breakthrough, a peer review may very well judge it to be absurd and of no values. As noted by Van Flandern (1993, ch. 21), peer review in journals interferes with the objective examination of extraordinary ideas on their merits. Maddox (1993), who was editor of the journal *Nature*, has said that if Newton submitted his theory of gravity to a journal today, it would almost certainly be rejected as being too preposterous to believe. On the one hand, there is a failure to select novel ideas (Brezis, 2007; Horrobin, 1990). On the other hand, the refereeing process trends to conformity (Mahoney, 1977; Martin, 1997, ch. 5). When one writes a paper that repeats what has already been said by hundreds of other papers on the same topic—with some changes, perhaps, in the parameters if using a theoretical model, or focusing on different objects than those which have already been analyzed, or analyzing the same objects with better data—and reaching conclusions which are already well known and in agreement with everybody's views (especially those of the referee, who is usually a mediocre representative of orthodox ideas; an exception might occur in relation to some secondary points), in these cases, a referee will be more likely to be less belligerent and may even send congratulations to the authors. Scientists are educated nowadays in a habit of self-censorship. The system promotes self-repression in the spread of ideas, so most scientists, when writing a paper, think something like "I think this and that, but I cannot say so in my paper because this will not pass the referee's control, so I will not say it". This causes serious harm to creativity among people who dare to think new things. Fortunately, there is always somebody who dodges this self-repression and wants to think freely.[1]

There is plenty of evidence of bias in support of papers confirming the currently accepted viewpoint and in favour of established researchers (Armstrong, 1997). Evidence for existence of bias for and against nationality, language, speciality, reputation and gender have been reported in the refereeing process (Wenneras & Wold, 1997; Godlee *et al.*, 1998). Nepotism (friendship networks) (Wenneras & Wold, 1997; Thurner & Hanel, 2010) is also common.

Reputation is an important bias. When the author is a reputable researcher, the work is usually given more consideration than when the author is not well known. In an experimental study, Peters & Ceci (1982) selected twelve previously published research articles by investigators from prestigious American psychology departments. After slightly altering the articles and changing the author's' names and their affiliations, Peters & Ceci resubmitted the manuscripts to the journals which had originally refereed and published them 18-32 months earlier. Only three of the resubmissions were detected, and eight out of the other nine were rejected. The reasons for rejection in many cases were "serious methodological flaws".

There are many examples of papers unfairly rejected by journals. Nobel Prize-winner Subrahmanyan Chandrasekhar produced a theory of stellar development which was not published because of strong opposition by the establishment. Twenty years later the theory was developed again and published by others (Wali, 1991, ch. 6). The

---

[1] By the way, this book *The Twilight of the Scientific Age* contains plenty of polemical points of view, which might be supposed an obstacle to its publication. Possibly I could be less belligerent, if I only wanted to publish a book in order to benefit my curriculum. However, criticizing a system which represses the spread of some ideas for the sake of surviving in the system and getting a job would not be consistent with the self-repression of my text. I may be wrong in many points, but the style of this book is consistent with its content: a claim for free thinking.

biologist Lynn Margulis is today recognized for her theory of the origin of eukaryotic cells as a consequence of the symbiotic union among different prokaryotic cells, which was published only after being rejected by about fifteen journals (Brockman, 1995, p. 135).

The peer review process is flawed in many aspects; it presents many problems, and there is no evidence that it works efficiently to select the best science (Thurner & Hanel, 2010; Bauer, 2012). The background problem is as follows: referees are people who have dedicated their whole life to doing research on the few problems of a particular field. They are usually people widely recognized in their field and their social status is due to their contributions in the field. The referees may be young postdocs, too, rather than researchers with a higher status, but in some degree their recognition also comes from their contribution to established ideas. As people with experience and prestige, sometimes accompanied by an excess of vanity, they usually think along the lines of "I am a great specialist in this field. I know the interesting and crucial ideas about it. If a new idea were presented, either it is wrong, it is of little interest, or I would have thought of it before. Therefore, if somebody presents a new work that tries to tap into crucial questions, either it is a continuation of my own work and ideas and those in which I was involved, or it is wrong". Moreover, it might be misconstrued as competitive, that is, argumentative or contrary, for somebody to publish a theory or interpretation different to that held by the referee. Perhaps it is somewhat exaggerated to attribute this thought to many of the authorities in a field. However, I think that something like this thought is often present. Of course, this vanity is not explicitly recognized. The fact is, this psychological mechanism, although not explicit, is perhaps present in most of the cases in which there is a discussion about the credibility of a theory, whether it is convincing or not, or any other subjective approach. Certainly, science has an objective content, and the data and maths are there, independent of what is thought about them. However, the data interpretation and the plausibility of a theory is something that is subject to the human factor—beliefs and, in many cases, prejudices. This carries much weight in the censorship of scientific publications. Of course, I must also say that there are many good referees who do wonderful work.

It is not only that interesting ideas are rejected, but the authors of these ideas are also demoralized and discouraged from continuing the development and/or improvement of their ideas. The way the system rejects new ideas is repugnant. The usual method for the editor of a journal is to try to give the impression that there is a general consensus that something is wrong. It is customary, for instance, that when a negative report on a paper comes from a referee, the paper is submitted to a second referee (this makes the evaluation process sound quite honest), but the report of the first referee is also sent to the second referee (which is not so honest if one wants an objective evaluation). Thus, the second referee reads the report of the first referee, then they have a quick look at the paper, and usually says "I agree with the first referee report". With this, the editor claims that they have sent the paper to two independent referees, that the editors have carefully evaluated the paper, and that all of them agree that the paper is flawed for the same reasons. This happens very frequently. In my experience, when a paper rejected with this kind of editorial consensus is submitted to another journal, the consensus disappears and one finds totally different objections to the paper, or else acceptance.

What is the consequence of all this? It really has a positive effect: avoiding the publication of hundreds or thousands of incorrect papers with absurd ideas that make no sense. Peer review is a process useful for giving science credibility, giving some value to people's careers, preserving the quality of research. Nevertheless, the negative side is also obvious: obstruction of thought and the few interesting ideas that might be produced. Work done well but without ideas, the work of specialist artisans, is rewarded. Creativity is blocked. It seems that the system gives the message that no ideas are needed. It seems the system, through its higher authorities, is saying that science only needs to work out the details. It is accepted that the basis of what is now known is correct, that present-day theories are more or less correct and only manpower is needed to sort out some parameters or aspects of minor importance. A Copernican revolution is totally unthinkable within the current system, even if the truth turns out to be different to present theories. With regard to this, there are not many differences between the modern academy and the universities in the sixteenth and seventeenth centuries, which conformed to the church and to Aristotle's texts. The consequence of it is that peer review tends to concentrate funding on the most popular and mainstream topics (Gillies, 2008, ch. 3), whereas it is inimical to new science (Braben, 2004). Frey (2003) regards this as a form of academic prostitution in which scientists have to choose between developing their own ideas freely or being constrained by subjects which allow academic success. Marvin Herndon (2008) even thinks that it is the basic cause of corruption in science.

Do you think I am too pessimistic about how the system of evaluation in journals works? There are people who are even more pessimistic than me. Taschner (2007), talking about the peer review system, says: "...the illusion that papers written by researchers are really read by those colleagues who keep the power of important decisions. In my view, the situation—at least in some disciplines—is much more miserable: often nothing is read any more, but, in the best cases, good friends among the gatekeepers are asked by phone or email whether the author is really suitable".

There are other ways to publish results, mainly on the web or preprint servers, but there are also filters and the way to achieve recognition with these unofficial publications is also very limited. Somebody might also steal a researcher's ideas, but that also happens with papers accepted in reputed journals that went unnoticed at the time of publication until years later a prestigious author rediscovers them, and picks up on the ideas. How many authors of the old Soviet Union discovered interesting things, which the world did not know about until a clever North American researcher, with plenty of dollars, rediscovered their work.

The most important tool for communicating scientific results in physics is the preprint server *arXiv.org*. It is a monopoly within physics and has no competitors. Even most of the papers published journals are posted on this preprint server, and people read them here. The situation is that papers not posted on *arXiv.org*, will receive scant dissemination within the community, particularly when the papers are not published in a reputed refereed journal, which is often the case for non-mainstream positions.

The development of *arXiv.org*, first at Los Alamos National Laboratory and later at Cornell University, was a wonderful example of freedom of expression between 1992 to 2004 that provided everybody with an open forum in which to post their ideas. There was a small fraction of papers with exotic ideas, but they were very few (5% or less), so they did not disturb the flow of information. However, after 2004 there was a change in policy and those responsible for the site decided to block the posting of certain contributions. In 2004, a system was introduced in which in order to post something on the site support was requested from a colleague with experience in the field. The system would become more perverse in the following years, forbidding some scientists from giving support when arXiv moderators noted that they had allowed the publication of very challenging heterodox ideas, and creating committees to reject papers without having read them and with the absence of a referee's report: the committees just read the title and the abstract and, if they did not like the content (and normally they do not like anything that smells of heterodox ideas, they channel the paper from the specialized section to their section of general physics 'physics.gen-ph', which is hardly read by anybody. In some cases, they remove the contribution totally, without further explanation (e.g., Castro Perelman, 2008). When asked for an explanation for a rejection, they usually reply with set phrases: "arXiv reserves the right to reclassify or reject any submission. We are not obligated to provide substantive reasons for every rejection, and usually the moderators do not provide more than a sentence or two, often in a form not appropriate for author viewing". This method of censorship of the promotion of new ideas is on a par with censorship in the Middle Age or in certain totalitarian regimes.

Actually, the main problem is not direct censorship itself, but the screening action of the massive overproduction of papers, with millions of scientists producing millions of papers every year, the reading of which cannot be undertaken by even the most hardworking of readers. According to some philosophers of science, every fifteen years approximately, the mass of scientific knowledge is doubled (Iradier, 2009), which means that in a century the production of scientific results is multiplied by a factor around one hundred, resulting in the multiple creation of subfields within a field, and sub-subfields within a subfield, etc. Even within a microspeciality, the number of papers may be around a thousand per year, still a huge amount even to take a quick look at. This number continues to grow in an uncontrolled way. The Indian physicist Subrahmanyan Chandrasekhar, one of the old editors of *Astrophysical Journal*, after leaving his duties as editor, recognized the increasing flow in the number of papers per year. He used to say, ironically, that the increasing velocity of the number of papers was greater than the speed of light, but there was nothing to worry about for there was no violation of any physical law because these papers carried no information. Schroer (2011) has said, referring to the field of particle physics: "Clearly the aim behind such mass-publications was not to clarify a scientific problem, but rather to be on the side of a career-supporting trend leading possibly (at least for some) to grants, fame and prizes". Since most of these papers do not contribute anything important to the field, only dispensable details, the odds of the few important papers which undergo censorship that would otherwise have an impact on the community are significantly reduced. This means that, once the obstacle of direct censorship in the journals is removed, the re searcher who tries out new ideas will have to fight with indirect censorship: the super-production of papers that conceal what is not of interest to the system. Propaganda is the key element in a paper becoming known. For this, the leading specialists again have the advantage, because they control most of the strings which move the publicity machinery; they have the appropriate contacts, they write reviews (summaries of scientific discoveries within a field), they organize congresses and give talks as invited speakers. Moreover, the reproduction of standard ideas is more acceptable because many people are interested in them, while the diffusion of new ideas is of interest only to their creators. This is not something new, it has happened all throughout history. The new thing is the institutionalization and bureaucratization of this process.

3.4   Scientometrics

Among the many perversions of the present-day system is the creation of a science of measuring "quantitatively" the advances of science. This is called scientometrics, and basically it involves bibliometric measurement of the citations of scientific publications. This quantification is needed to justify the distribution of funds and positions.

In my opinion, evaluating the quality of science by the number of citations it receives is like evaluating the quality of a TV program by its audience: the most popular works are not usually the best ones but the most mediocre ones. The concept that the best ideas are those which any idiot can understand and repeat is a democratic one. But nature is aristocratic, and the major advancements are only appreciated by a minority intellectual elite. Of course, once they are accepted by the establishment, once several generations have passed, even the fools can laud them.

The factors which contribute to the major impact of a scientific work are diverse. In principle, they measure the popularity of a paper but also reflect the number of scientists working in a specific field (Habing, 2009). And the factors will depend also on the specific subfield. In astrophysics,[2] the number of citations depends statistically on: the journal in which the paper is published; the visibility of the paper in preprint servers and its position in the lists of preprints (Dietrich, 2008a,b); the length of the paper (Stanek, 2008); the subject matter, instruments used and their wavelength (Trimble & Ceja, 2008); etc. And, of course, major factors will include the publicity surrounding the paper and the status of its authors. There are also citation clubs in which researchers do deals to cite each other's papers (Gillies, 2008, ch. 6). The leading science administrators (see subsection 3.6) make greater efforts to publicize their mediocre scientific results than to create brilliant scientific results. There are also people who join big research groups but participate little yet get a CV with a huge number of citations. There are even cases of individuals who have never written a scientific paper as first author yet are among the most cited scientists in their speciality.[3] Difficult papers are less frequently cited, because most scientists are not willing to spend time with them, and they just read the abstracts of some papers from which they can pick a number or sentence which is easy to cite. All these elements and many more carry a greater weight. It is normal for things to be this way, and it has happened throughout history that fame was unfairly distributed. What is not so normal is the creation of a system of evaluation in which this unfair distribution is used as a measure of quality.

Certainly, there are reasons for the existence of this naïve system of evaluation. And this has to do precisely with the topic of this book: the meaning of scientific research. In a time when the scientific discoveries were of the first order, it was not necessary to invent a system to distinguish between good and bad science. It was evident what was relevant or not. However, at present, among a huge quantity of scientific results of second and higher order types, among the huge amounts of data and publications and persons moving from one place to another, it is difficult to see anything very important. How can one distinguish the good science from the bad science

---

[2] I often use the example of astrophysics in this book and I give more statistics about it because I have more information for this than for other disciplines. In any case, the arguments given are general for all sciences.

[3] For instance, the 4th ranked astrophysicist by citations over the decade of 1997-2006 has never written a paper for a refereed journal as first author, and has gained almost all his citations through his right to sign official papers for a large collaborative group in which he played only a functional role (White, 2007).

when practically everything is the same, with no obviously revolutionary or important ideas? It is not evident. And this uncertainty about "good" science produces fear: fear of nihilism, of an empty enterprise which can no longer produce great advances. People need to have values to believe in; they need to believe that there are indeed ways to distinguish between the good and the bad science, and that good and important scientific results are being produced by our civilization.

The same thing happens with awards such as the Nobel Prizes or similar things. It also happens in other areas of culture. For example, in literature, people tend to believe that there is a good literature and bad literature, and that the many prizes and marketing operations for the creation of bestsellers separate them. The fact is that we live in an era with rubbish everywhere, and texts with nothing new to say but simply repeating the same stories for commercial purposes, far from the bright creativity and art of classical literature, but we try to convince ourselves that there is still something important to read among work produced by our contemporaries. People want to avoid being sceptical, and want to believe in the dogma that the best writers win prizes and are bestsellers. Thinking about the economic/power/status motives behind most cultural movements is too hard for ordinary people. If they were to cast doubt on the system of evaluation, they would lose their way. Therefore, evaluation criteria are necessary in present-day science; and they have to be objective, or have to produce the impression that they are objective. The number of times a paper is cited was chosen from among the possible criteria as the dogma of those which believe in the system, perhaps because it is simpler to measure than other methods which require further thinking about science;[4] we need only count, and this can be carried out even by a person who does not understand a word about science, an ideal method for bureaucrats of science.

---

[4] For instance, Knuteson (2007) proposed that the value of a particular experimental result in an academic field should be measured by how much is learned from the result; as a counter, the value of a result should be how surprised you are that the particular result has been obtained. Knuteson (2009) proposes something similar for theoretical results.

3.5   Congresses

Congresses, symposia, workshops, schools, meetings, any opportunity of meeting other professionals to talk and exchange ideas, have become widespread. The phenomenon is not only present in the sciences but in every professional environment, and has increased hugely during recent years. In the first decades of the twentieth century, while discoveries of huge importance to the development of physics (for instance, in relativity, or quantum physics) were being made so rapidly, such gatherings occurred once in a blue moon, to celebrate important advances. Think, for instance, of the Fifth Solvay International Conference on Electrons and Photons, held in 1927, where the world's most notable physicists met to discuss the newly formulated quantum theory. Seventeen of the twenty-nine attendees were or became Nobel Prize-winners (and the Prize itself was not important, so much as the front-line discoveries which earned the Prize; nowadays, Nobel Prizes are given to discoverers of mostly secondary things). The First Solvay Conference was in 1911, and it was indeed the first world physics conference. Today, there are thousand of international congresses every year, in physics alone, quite apart from small local or national meetings, with hundreds of mediocre participants at each. There are even macro-meetings, with thousands of researchers. The saddest part of the situation is that the conceptual level of development of physics today is far below what was reached in the beginning of the twentieth century. In other areas with a much larger population of researchers, in medicine for instance, the numbers are even higher.

Holidays can be a reason to attend congresses. Many of them are held in exotic or tourist destinations, which allows leading scientists and their friends to enjoy a holiday using public funds. However, the main purpose of congresses is not to promote tourism but the diffusion of information in a micro-field of science, trying to give a broader, more global overview to a given topic. In order to do that, the congress is usually structured as a long series of talks that last several days. The invited speakers are highlighted and they are allowed to give long talks, of up to one hour, to speak about their own research or those papers in which they are interested. They comprise ten or twenty leading specialists who are friends of the congress organizers or share similar ideas. There are also selected speakers who are among those who apply to give a talk. Since there are so many, the available time is distributed so that each of them can speak twenty or fifteen minutes or even less. In this short time, they must discuss their research activities during the last two or three years. Consequently, the result is concentrated talk sessions that quickly exhaust the attention of the audience. Basically, they consist of propaganda. It is useful to say that I have carried out work on this, and that anyone who wants to know about it must read my paper.

Finally, there is a room for posters, in which hundreds of abridged "scrolls" concentrate texts and figures onto a square meter of bristol board per poster to show results and gain propaganda value from them. This reminds me of the trade exhibitions at which each company shows its merchandise for publicity. Moreover, publicity resources to attract the attention of the congress attendees (the same people who show posters or present papers) with pictures and colourful poster designs, videos of numerous simulations or films that make an impact on conversation (there are even cases of researchers who pay professional animators to make the videos), etc. All this has the goal of attracting the attention of an audience that is lost among the tons of information; dispensable information since there is little that is new to say at each congress, just simple technical details without too much relevance. The battle of the scientist is not in finding good new ideas, but in finding the way to sell mean, unworthy ideas. Marketing is more important than scientific tools. It is all just publicity, and meeting colleagues to talk about, and discuss future collaborations.

3.6 Financing, scientists transformed into administrators/politicians of science, and "supervedettes"

Many senior scientists and functionaries with security of tenure devote most of their time to teaching at universities. Perhaps they take a "slave"—I mean a Ph.D. student—to carry out work which the senior scientist will then co-sign, in order to show that they actually do some research. In first-rank institutes, competition is higher. In those institutes, some of the researchers are project leaders, and pursue "impact"; that is, the project has the objective of producing many published papers, and they command a certain level of respect from specialists.

The project's main researcher is the leader of a group comprising several Ph.D. students, several postdocs and, perhaps, some senior scientists of lower status. There are even cases in which this main researcher may have all the postdocs of a small institute under his control. This main researcher is usually a kind of commercial manager, and in some cases could be termed agent and adviser.

They begin their careers as scientists, but they become administrators or politicians of science. In astrophysics, I used to call them "astropoliticians". One may ask why scientists become mere administrators. Isn't research more fun than administration? The answer to this question is again related to the nonsense which science has become. Indeed, these administrators are often victims of the system rather than being responsible of it. A scientist who has spent ten to twenty years doing research realizes that most likely he will not make any revolutionary scientific discovery, that scientific tasks are not related to new ideas but to technical applications of old ideas. The few intelligent ideas they may have are likely to be ignored among tons of written papers. The real science of interest nowadays is the science accompanied by ballyhoo; by being a science administrator one get more chances to be involved in it and a higher status in the priesthood. The situation is well described by Gillies (2008, ch. 8): "Academics typically start with great enthusiasm for research, but, after a number of years working at research, they often become rather bored with it. They may have run out of ideas. They may have come to realise that their youthful hopes of becoming the next Einstein were an illusion, while the reality is that there are quite a number of young researchers doing better than they are. In these circumstances the sensible move is into administration and management where a tempting career ladder stretches before them".

Some researchers claim that they have so many bright ideas that they cannot develop all of them, thus they need an army of students and postdocs to develop them, and that is the reason for becoming a leader of a project and administering the research time of young scientists. Apart from the arrogance implicit in that position, we must think that if the ideas were so good, they would not be left to low-ranking workers. Only massive industrial ideas need a lot of people; artist's ideas are better developed by the artist, not by sub-workers.

I think that most administrators exhibit a similar behavioural pattern. One must make an appointment to simply talk with them, since they are always busy with a thousand and one tasks. "I have no time" is one of the favourite sentences of these businessmen, men of our era. We live at a time in which even pipsqueaks pretend to conduct themselves as if they were important (a minister or someone who is really important) and deliver the self-important response, "I have no time" or "I am busy".

In their offices, it is usual for them to receive three or four phone calls in less than thirty minutes. They receive hundreds of emails daily. When an appointment is required in order to present some scientific results for an opinion, the administrator has to revise their agenda, mentally or in a notebook, because there is always a meeting to be attended somewhere. In addition, much travel is undertaken, both nationally and to foreign countries. They must prepare talks, because they are the main speakers at various congresses. They must attend a large number of meetings with other administrators to make agreements (scientific collaborations, not commercial agreements, but the outcomes are similar), or negotiate budgets, or create propaganda for the project in order to obtain some economic benefit or achieve an impact in some other way, or to think up—together with other administrators—yet another macro-project that will cost many millions of Euros and will employ many researchers in yet more monotonous work. Of course, they are not those researchers but people who follow their orders, along with other new slaves who will be brought on stream with the money received from various negotiations. When they are not travelling or in meetings, they usually are busy with the preparation of periodic information bulletins concerning project activities or filling in forms to apply for the use of instruments or applying pressure for economic support for the project (travel, computers, scientific instruments, etc.) to some ministry or other for various types of assistance. They also spend a lot of time thinking about how to spend the funds obtained for their research, because a scientist who spends modest amounts on travel, computers, etc., is not considered a good scientist. Giving back part of the funds is not considered a good sign in a scientist. A good administrator/politician of science must ask for a lot of money and must spend whatever they get (possibly changing computers or hardware every few years or buying unnecessary things for the department), in order to give the impression to their colleagues that "high level" science is being carried out and that any amount of money is insufficient for the huge needs of such impressive science.

In their spare time, they usually are busy with the coordination of project staff and their work efforts as well as establishing research priorities. The hen takes a cup of coffee to rest from bureaucratic duties, while the little chickens are all around, eager to show their results. The administrators listen to (in many cases, they do not listen), and read papers by low-ranking "workers", express their opinion and, more than likely, suggest changes, according to their prejudices. It is a rare day when they sit down to do some actual scientific work, in the true sense of the word "scientific". Perhaps, they may dedicate a few hours some days to teach something to a low-ranking science worker. In most cases, however, it is not they who dedicate months to work on resolving various problems but, instead, their Ph.D. students or postdocs.

When an administrator or politician of science is able to display particularly sparkling qualities as an agent and trader for the science that created these workers, we have an example of a "star": a star that is dazzling in both its brilliance and prominence. In other words, a *supervedette*, the great star among the stars. In an institute with more than a hundred researchers, there are usually only one or two *supervedettes*. It is not difficult to identify them because they are essential reference points for the particular institute, especially in the image presented to the world outside. If a journalist visits in order to write an article about what is happening at the institute, the *supervedette* comes to the fore. If their team does any work, the press is quickly called in to announce to the world what so-and-so "et al." are doing; that is, "and collaborators", although the names of the low-rank workers who have made the discovery is usually not important, fame being focused on the *supervedette*. They have usually a good eye for choosing topics that have high popular impact (not necessarily topics of high scientific importance). If the topic is unproblematic, they will announce it with a lot of ballyhoo in order to create a fuss. They have no difficulty publishing in journals; they write professional and popular books. Together with others of similar ilk, they own the congresses, they have access to expensive instruments, and the budget for their activities is gargantuan. They think in terms of major not minor goals. They lead large multimillion-Euro projects. In addition, a single phone call can translate into widely disseminated propaganda via journals and television. They compete as "the best researcher" for domestic and international grants. They mix with people in high society. Circumstances may vary depending on how big the star is, but they are—in short—a subject of envy for any administrator.

Of course, there are exceptions to this behaviour. Any attempt at generalizing a behaviour pattern of a given group is always subject to correction in individual cases. There are some instances in which a senior scientist, even the leader of a project, works at the same tasks as lower-rank workers, and does not spend too much time on administrative tasks. This, however, is not the usual situation. Nowadays (and most likely in the past as well), the "trumpeter" is another species of scientist: the professional science agent, the executive with attaché case in hand. In many cases, this agent does not know how to approach solving a scientific problem, nor does he have too much understanding of the appropriate scientific field. From an objective standpoint, many of the discoveries made by these administrators do not deserve to be highlighted, but this objectivity is difficult to achieve when all researchers believe their own works are important. The kingdom of bureaucracy triumphs.

3.7   Press, television, propaganda

As mentioned earlier, press, radio, television or other media, are useful tools for the manipulation of information and commercial propaganda. Our knowledge of society in relation to scientific activities in general stems almost totally from press, television and propaganda sources. Therefore, manipulating these media is an ideal way of achieving what the controllers of a society desired to have perceived or believed by the general public.

Most journalists responsible for writing articles about science have little knowledge of what they are writing about; perhaps they have some general scientific knowledge, but they are very far from having a grasp of all the existing specialities. This is the situation in even the most prestigious newspapers. Among the less prestigious ones, it is even more likely that their journalists will have no scientific knowledge whatsoever. Because of this, journalists are obliged to believe what the researcher says. If they are told that a high-impact discovery has just been made, the journalists must trust that it is so, since the journalist has no personal knowledge from which to cast doubt on the veracity of the researcher's statements. The determining factor in these situations is the researcher's reputation. Thus, fame feeds fame: a prestigious researcher is usually surrounded by a swarm of journalists. The propaganda they distribute will contribute to increasing the researcher's "fame". In this respect, there are not many differences between the "fame" achieved by a scientist and that which a singer attracts: it is all a question of being available to the mass media.

Researchers perhaps overestimate the value of their own work, but do not usually distort or exaggerate, or say that things are the opposite of what they are—at least not intentionally. Journalists do all these things and do them intentionally. The goal is impact, which is something of high value among friends of misinformation and ballyhoo. It is, apparently, what they are taught in the faculties of journalism. Thus, a good deal of the information published in the press about recent scientific discoveries contains significant errors and receives appraisals which are totally inconsistent with the purported newsworthiness of the reported item. Titles and headlines often distort the news. I still remember reading in a newspaper something along the lines of "extraterrestrial mummy", in reference to the fact that some researchers had found the tomb of an ancient Egyptian pharaoh that had been built with stones from a place where a meteorite had landed on earth in the past.

The number of cases where scientific news is published with a disproportionately huge amount of commotion—for example, that Einstein's theory of relativity is no longer true, or that there is life on Mars, or that cold fusion works—is very high. In many cases it stems from the misinterpretations made by journalists because they do not understand the subject. In other cases, the sources may be genuine discoveries that have been published in scientific journals, but which are still being discussed and about which certain controversies remain. After some months of sensationalistic reporting, the scientific community usually clarifies that the discovery was not really a discovery because there were some errors in their results. However, the general public only remembers the huge kerfuffle, not the reply that refutes it. Apparently, for commercial purposes the truth is not so interesting, and does not help to sell more newspapers or magazines. If somebody wants to know about science, I would not advise trying to find out about it through the press or television, but through textbooks. I would also advise people to forget the newspapers. The future will tell us what is being done properly now.

In spite of the imperfections of scientific communication throughout the mass media, it remains the fundamental pillar of the relationship between scientists and society. Many of the multimillion-Euro grants depend on it. For example, the case of the Antarctic stone containing life of Martian origins was famous all around the world. It gave rise to a large grant for further research into the subject from the U.S government. Afterwards, the news was denied—the stone was contaminated with terrestrial life—but those who got the money for the project had already obtained what they wanted. Incidentally, the paper about this discovery was published by *Nature*, a professional journal of prestige. The paper itself had been submitted to three or four referees and accepted.

In some cases, the opposite thing happens. Funding is not a consequence of the press; rather the press is a consequence of funding. When large amounts of money are invested in a project, justifying the investment of public funds becomes necessary. They say, for instance, that a satellite or a telescope is going to create a revolution in astronomy, that we are going to observe the whole Universe and parts of other ones. This, again, is propaganda. In some cases there are moderately important discoveries, but in many cases there is nothing interesting. The instrument arrives, but the revolution has been rather small. Perhaps they scrape together something else about some galaxy which was not in the study. Specifically, in the last examples the press is needed to exaggerate and create the belief that an item's newsworthiness is greater than it actually was. The press is usually called in to explain the great discoveries made, thanks to the taxes paid by the nation.

Moreover, the press is not always at the service of all-important scientific phenomena. Without fame, without money and without the recommendation of, or support from, a prestigious team of researchers, even the best of scientists, working in the most important fields, would be not listened to.

"An individual with few resources, achieving what we could not do with billions of Euros. This is a scandal, and we cannot allow it". This is the message of the real capitalist society where money exerts its power. A new Einstein working in a patent office would be a scandal.

3.8 Snowball effect in the dynamics of social groups in science

The snowball effect, also called the Matthew[5] effect (Merton, 1968), is present to a certain extent in the social dynamics of science, especially in the most speculative areas. It is a feedback loop: the more

---

[5] Merton (1968) gave it the name "Matthew effect" from the Gospel of Matthew (25:29) which says: "For everyone who has will be given more and he will have an abundance. Whoever does not have, even what he has will be taken from him".

successful a line of research is, the more money and scientists are dedicated to working on it, and the greater the number of experiments on observations that can be explained ad hoc, such as in Ptolemaic geocentric astronomy; this leads to the theory being considered more successful.

In some cases, the system supports conservative views, but there are also cases of speculative lines of research that have been converted into large enterprises. For instance, in theoretical physics, string theory has absorbed a lot of people and funds, as well as marginalising and deprecating other approaches to the same problems (Luminet, 2008). Lee Smolin (2006) thinks that string theory is not only speculative but the conclusions are circular, the concepts are arbitrary and the hierarchical structure of this scientific community is quite outlandish. The Nobel Prize winner in Physics, Sheldon Glashow, wonders whether string theory is not more appropriate for an Institute of Mathematics or even a Faculty of Theology rather than to an Institute of Physics (Unzicker 2010, ch. 14). Unzicker (2010, ch. 14) considers physicists working in that theory as being like a sect or mafia. Another case is the search for supersymmetric particles in dark matter, which occupies more than thousand people at CERN. And what happens when, after a long period of search, when huge amounts of money have been consumed, the experiments or observations do not find any evidence in favour of these theories? Then the groups claim that we must carry out exploration at higher energies and they ask more money

A general trend in the massive overproduction of scientific results is the listing of a large number of authors on scientific papers, with the number of single-author papers significantly reduced. This has been happening in the last few decades and the situation is changing quickly. For instance, in astrophysics, in 1975, 40% of the papers had a single author and fewer than 3% had six or more authors; in 2006, 9% of the papers had a single author and 28% had six or more authors (White, 2007). These and other statistics show us that science is becoming an enterprise of big numbers (number of people, amount of money...) like multinational industries, in which the role of creative and independent individuals thinking about science is being substituted with corporate democratic-capitalist consensus science. Unzicker (2010, ch. 1) thinks that in physics there are no longer idealistic individual thinkers, only large organizations, political interests and the rules of the science market. The role of the genius is being filled by the average technocrat. Like a snowball rolling down in a hillside, or like huge black holes drawing in surrounding matter, or like large enterprises absorbing or eliminating small companies in a capitalist economy, the big groups in science absorb the resources, and the small groups or individuals are forced to join these big groups if they want to survive, otherwise they are ignored or pushed out of the system. Sometimes the joining together of small groups to form a big one is just a way of getting funds from the government, since it is easier to get millions of Euros for a macro-project in which tens or hundreds researchers work than to get a modest amount of a few thousand Euros for a small group of researchers, even though the collaboration may not be genuine. Apparently, the projects of macro-groups instil higher confidence in funding bodies, as if to say "so many people working together should not take a wrong direction in their research". However, these macro-projects are frequently macro-fiascos.

The important thing in science is ideas, and only individuals can think. Groups do not think; they just follow a mechanism of social division of work. There is no such thing as the collective creation of an idea. In science, as we have seen in Chapter 2, all important ideas were created by individual geniuses. There was never an important achievement made by a big team. Only minor secondary details which require large amounts of manpower and small amounts of brain can be dealt with successfully by large teams. Technical details belong to collectives. Ideas belong to individuals. The brain is something individual. There is not such thing such as a collective brain.

"Man cannot survive except through his mind. He comes on earth unarmed. His brain is his only weapon. Animals obtain food by force. Man has no claws, no fangs, no horns, no great strength of muscle. He must plant his food or hunt it. To plant, he needs a process of thought. To hunt, he needs weapons, and to make weapons—a process of thought. From this simplest necessity to the highest religious abstraction, from the wheel to the skyscraper, everything we are and everything we have comes from a single attribute of man—the function of his reasoning mind.

But the mind is an attribute of the individual. There is no such thing as a collective brain. There is no such thing as a collective thought. An agreement reached by a group of men is only a compromise or an average drawn upon many individual thoughts. It is a secondary consequence. The primary act—the process of reason—must be performed by each man alone. We can divide a meal among many men. We cannot digest it in a collective stomach. No man can use his brain to think for another. All the functions of body and spirit are private. They cannot be shared or transferred". (Ayn Rand, *The Fountainhead* [novel], 1943)

As noted by my friend and colleague Eduardo Battaner (2006), human intelligence works in the opposite way to the intelligence of ants. Ants are not very intelligent when they are isolated from one another; however, in a group, they become more intelligent, and they organize their lives in a very clever way. Humans are individually intelligent, but when they associate in big groups this intelligence is diluted and transformed into macro-stupid behaviour. They become a mass, and a mass does not think but follows its leaders blindly, or else they simply organize their lives in terms of simple socioeconomic rules such as the capitalist way of living. As said prophetically by the Spanish philosopher Ortega y Gasset in his work *The revolt of the masses* (*La rebelión de las masas*), the twentieth century was the century of the masses, and it continues to be true in the twenty-first century. Masses of people transform religion into fanaticism rather than an intimate spiritual experience. Masses convert the state into machinery without sense of justice; even horrible episodes such as the resolution of the "Jewish problem" by the Nazi State in Germany during the Second World War can emerge from a state composed of automaton following orders. One may ask how a whole country composed of individuals can be so criminal as to kill hundreds of thousands of people in the name of purifying the race? The answer is that separately these individuals rarely behave so, but when they are integrated into a large group (sect, political party or whatever) they lose their values and become a blind mass directed by leaders, such as Hitler. The Spanish Inquisition, Stalinism, Neoliberalism, etc. are other examples. Masses also convert culture into a dehumanized thing, cold machinery without passions. Passions are necessary in order to feel the life embedded in culture. Science, as part of culture, as part of human expression, should also be a passion, as it was throughout history, with many individuals feeling the emotion of scientific creation. Elevated passions are proper to individuals; masses as a whole do not feel them.

"In science, old boy, discipline is not at all useful. Work with the motor of passion and with the brake of objectivity. Leave discipline alone. Discipline is good for the militia; tenacity is good for religiosity; passion is the only thing which moves science". (Battaner, 2010, *El astrónomo y el templario* [novel] [*The Astronomer and the Knight Templar*])

4 KNOWLEDGE

The fruit of the tree of science is called knowledge. This fruit is comprised of abstract things whose value cannot be directly perceived. Perhaps the parameter of the value of knowledge should be its "importance", but its measurement is very subjective. There are attempts to quantify the importance of particular knowledge, as mentioned in the previous chapter, referring to scientometrics, but, as also noted, these are simple naïve rules which do not touch the true essence of what represents an important discovery. Indeed, determining which knowledge is important is as difficult as evaluating whether a poem is good; one has to be as clever as the creator in order to be able to evaluate the merits of his/her creation, but in most cases scientists are just ordinary people, a long way from being geniuses.

Another remarkable aspect of knowledge nowadays is its rapid growth. Nature appears to be an unlimited source of phenomena, with each answer followed by many new questions needing to be solved. Of course, not all the questions are equally important and usually the initial questions are more relevant than the ones which follow, but since the system of evaluation of the "importance of knowledge" has become so relativistic, there is no brake in the scientific machinery, which tends to produce more and more. This gives us a scenario in which our stores are saturated with scientific fruits, and their market price (I mean the distribution of funds for the different lines of research) is regulated in a speculative way. Science has become another tentacle of capitalism, and the ideals of the search of truth occupy a secondary role within the system. Art is another case of human activity being converted into a speculative market, separate from the idea of beauty or artistic value. Perhaps economics is the science which has absorbed the rest of them.

Wondering about the significance of scientific research means asking about the importance of knowledge in our lives. What use does knowledge serve? And, given the amount of knowledge we already have about nature, are the great efforts devoted to funding science worth it?

4.1 Knowledge society

We are told that we live in a knowledge-based society, a feature of civilization in the developed world which distinguishes it from the barbarians without culture, without education. A knowledge-based society is associated with progress, with success in the quickly changing economic and political dynamics of the modern world. Knowledge is associated with innovation and development, with welfare, with the state of superiority of Western culture over other primitive societies. We are even told that underdeveloped countries are in danger of increasing their marginalization because of their lack of information, by comparison with the developed countries in a modern knowledge-driven society (e.g., Giddens, 2006, ch. 17). This kind of discourse justifies the frenetic activity within universities and research institutes. They contribute to increasing our knowledge. Our new society tends to substitute the industrial economy for a knowledge-based economy, where most of the jobs are associated with the administration of information (Giddens, 2006, ch. 18). And we always need more knowledge; we cannot stop pursuing it. Otherwise, we would lose our advantage in the race of civilization. This is not my opinion, but what seems to be inside the narrow minds of most of the science bureaucrats and administrators throughout the world.

I agree that some knowledge is useful and good for human beings. Indeed, I agree with the intentions of the pro-knowledge discourse. However, I disagree with the boundless limits of this pursuit of knowledge. I even think that this excess of knowledge is bad. People get lost among such huge amounts of information, and consequently they lose their ability to think lucidly. Knowledge, that is, in the sense of an amount of information, is not the same thing as wisdom. An encyclopaedia holds a lot of information, but there is no wisdom in it, just collections of data. There are some abnormal people, those are autistic, who possess extraordinary memories and are able to absorb information from hundreds of books, which they can repeat when asked. They also can show extraordinary abilities in arithmetical calculations. But they are not clever, wise, intelligent people; they are indeed slightly stupid. The same thing can be said about our society: certainly, there is plenty of information, plenty of data, but in fact knowledge is not very useful to society because it usually behaves in a quite stupid way.

Memory is not intelligence. Of course, we need some memory in order to put some information in our brain, but we do not need extraordinary memories. Rather, we might need more intelligence to solve our problems; for instance, the problem of a world which cannot stop contaminating and destroying the environment, or more intelligence to put the economy at the service of men and not men at the service of the economy. Science provides both things: encyclopaedic data to fill our memories, and brilliant ideas which are able to connect different phenomena and solve problems. Brilliant ideas are scarce, particularly at this time, but data and information are produced in exorbitant amounts. People sometimes confuse memory with intelligence, and think that further developments of science will solve all the problems of the world. Therefore, huge investments are made in order to carry out research in environmental problems, and … certainly we are more informed about the details of the fast pace of damage of the environment, and its causes. We have now got many measurements of temperatures throughout the world, with satellites monitoring these rapid changes because of global warming, for instance. But the situation in the world does not change because of this knowledge. Even worse, because of the huge amounts of money invested in environmental research, we get tens of thousands of people consuming further resources, using planes to travel to the many conferences they organize, and contributing along with the rest of the world to worsening the situation. Possibly, our society itself is becoming somewhat autistic, having access to plenty of information but unable to use it globally in an intelligent way.

It is also usually claimed that the mission of science and its associated scientific knowledge is to provide technological progress, useful for the welfare of people. Certainly, there is here a confusion between science and engineering. Again, I do not agree this kind of statements. Schrödinger (post-1996) thought that the mission of science was not technological progress but to contribute to wisdom; and he doubts that the happiness of human beings had being increased thanks to technical progress. So I think too.

4.2 Facing the limits of knowledge: on Horgan's position

The title of the present book, *The Twilight of the Scientific Age*, was taken from the subtitle of a book by John Horgan (1996): *The End of Science. Facing the Limits of Knowledge in the Twilight of the Scientific Age*. In my opinion, Horgan's book is of great value, and I find it an important reference in considering the subject of "the end of science". Nonetheless, its title and contents reflect an idea which I do not think is the key of the question. For my book, I chose the title *The Twilight of the Scientific Age*, removing deliberately *Facing the Limits of Knowledge*, because I think this is precisely the misguiding idea contained in Horgan's book: The intuition that scientific age is declining is prophetic, I guess, but not so the causes Horgan gives. Possibly the limits of knowledge is one of the causes, but there is much more behind the twilight of science, and I think it is more related to being sated with knowledge rather than to the limits of knowledge. It is more a sociological/anthropological question than a pure debate about whether there remain scientific problems to be solved. In my opinion, Horgan's book is brave, lucid, with plenty of good ideas on topics related to the limits of knowledge in science. However, what I miss in Horgan's book is a development of the theme "the twilight of the scientific age", distinct from those limits. Horgan interprets the end of science as being the end of interesting unknown things, as the limits to our ability to expand our knowledge significantly in important broad topics about nature, but he does not cast doubt on the pursuit of knowledge as a characteristic of human culture, whatever the limits of knowledge might be.

John Horgan is a scientific journalist, and a good one. He worked for a long time at *Scientific American*. As such, he has good contacts with the most famous and influential scientists alive (practically all of them now very old), and he tries to find the clues about the future of science by interviewing them. In my opinion, these old scientists, famous thanks to the mass media, do not, in general, offer fresh ideas for a new era in science (or the end of it). They still think with the mindset of an old, successful science, so they are not going to offer the best insights into the future of science, and its decline. Nonetheless, the fact that it is filled with these famous names has perhaps contributed to making Horgan's book a popular bestseller for the masses. Publication of books is nowadays a totally capitalist enterprise; publishing does not worry about revolutionary ideas but about benefits in the sale of a book. In my opinion, however, the fast track to the truth is only appropriate for a minority. Truth is aristocratic, not democratic. And I also believe that revolutionary ideas must be sought among intelligent thinkers who flee the hubbub of worldly life, not among *supervedettes* of the system.

Horgan is also North American, and as such has an Anglo-Saxon way of viewing the relationship between science and society. The people interviewed were also living in English-speaking countries. Hence, his book does not consider a different view, drawing on ideas coming from other cultures. In particular, he does not consider views such as that of Unamuno; this does not come within the Anglo Saxon world view. Horgan does not pay attention to the question of a society which might have become overfed with knowledge, even when there is still much to know; he would declare that position to be "anti-scientific". As with most people who have been surrounded by science for a long time, there is a failure to acknowledge a more irrational view of human beings, of their passions, of their loves. Instead, Horgan supposes that the love for knowledge is an obligation of our future civilization. According to him, we must want knowledge, and the only excuse for stopping the scientific industry is that we know already everything. But what about a society which does not wish to obey the priests of a "knowledge society"?

Anyway, let us go to the major statement of Horgan's book, with which I do agree. He claims that it is possible that pure (not applied) sciences will not yield more great revelations (such as evolution theory, quantum mechanics, the Big Bang theory), but only incremental diminishing returns. I agree, with some doubts about the Big Bang theory being a genuine great revelation, but in general I agree with his idea that the age of first-class discoveries and theories is over. We must not deceive ourselves. The more the history advances, the more difficult the achievement of a relevant truth becomes. Newton's scientific activities during one year of his life, carried out with a mere notebook and a pen, were more fruitful than the activities of thousands of the best present-day scientists during their entire lifetimes, using millions of Euros. It seems that there are many writings, much data ... but in the final analysis our comprehension of nature in global terms advances at a very slow, nearly imperceptible pace. Greater efforts bear less fruit.

"After the fundamental laws are discovered, physics will succumb to second-rate thinkers, that is, philosophers", says Horgan ironically (1996, ch. 3). Or: "Just as lovers begin talking about their relationship only when it sours, so will scientists become more self-conscious and doubtful as their efforts yield diminishing returns" (Horgan, 1996, ch. 9). There remain some major mysteries to be solved, he says, for instance in biology, concerning the origins of life, but the future research on these will not deliver major shifts in our vision of the world; and Horgan, thinking in a conservative way, does not believe that a major revolution will arise from the development of heterodox ideas to replace present orthodox ideas.

Certainly, I see the progress of science similarly, but most administrators of science and ordinary people alike are still living on dreams of an era of important discoveries, and it does not matter that sceptical views such Horgan's or mine do not agree with those illusions. The point is that there are strong economical interests surrounding the business of science, and the publicity in the mass media is used to create the illusion of great science, supported by huge investments of public money. The spirit of science is already in decline, close to death; but the body of science, the hierarchy around the business of research, is still very much alive and will not decline simply because of a lack of interesting results. The "interest of the results" is always a subjective, relative thing, and the people who earn money from the business of science will always defend the magnificence of their discoveries, with some honest exceptions; whereas the people who are at a distance from the science industry and dare to claim the emptiness of most scientific enterprises will be called "ignorant people".

An example of this magnified perspective of the importance of science was apparent in one of the special sessions at the General Assembly of the International Astronomical Union, in Rio de Janeiro (Brazil) in 2009. The title of this session was "Accelerating the Rate of Astronomical Discovery". Certainly, it was an attractive title, but my impression was the opposite of what they claimed: I would say that now we have a science which is becoming smaller and smaller, and more and more expensive, with more and more boring conferences (thousands of times more than in the 1920s), talking about ever more unimportant insignificant details for specialists. Some people, like Simon White, talked about a "big science"; they indeed meant science made with very expensive huge telescopes, satellites ... I say that "big science" should be a term for the science produced by Darwin, Maxwell, Einstein, quantum mechanics, the science, for instance, of the Solvay Conference, or the Great Debate in the 1920s. Now, there are some ideas which look like big mysteries of nature, unveiled by modern science (non-baryonic dark matter, dark energy), but at present these are just speculations, without strong support. We will see within thirty to forty years whether our science is big in results or just big in the astronomical size of the bill for the expenses; and whether the rate of "important" astronomical discoveries is increasing or we are just increasing the quantity of macro-fiascos and useless papers. The invited speakers at this special session were all of them mainstream researchers; there were no speakers going against the tide. Some authors offered some important criticisms while always emphasizing their politically correct opinion that the system works well, and that those criticisms are polishing the rough edges of a good system.

Probably, the end of Great Science is becoming more and more evident for those minds which are not sleeping or else being bribed by the system. But minor science has not ended. There are always some minor, unimportant, hidden phenomena which attract the attention of some scientists, in order to get funds for their research and ensure a *modus vivendi* with that excuse. Indeed, there is an established practice among scientists, inculcated since their early education, to never talk about the end of a line of research. It is politically correct to speak about the generation of new lines of research from an old one, a multiplication of areas, but not about the death of a line of research. When somebody talks about a final result, concluding that there is nothing else to develop from a particular line of research, they are used to being criticized by their colleagues as an arrogant person. More politically correct is the position of claiming that while some results have been obtained, further research is needed in this area, leaving room for a continuation; and it is even more politically correct if the scientist mentions that new expensive devices are necessary to develop that line of research further, requiring more further money, meaning that many new positions in science will be created. The great idea of an individual who tries to solve a problem once and forever, on his own, with few resources is usually considered as lacking a unity of interest with the wider community, and presumptuous. This behaviour feeds the impression among circles of scientists that there is always lots of science to do, with the end of science being very far away.

Indeed, the survival of the institutional body of science is not a matter of whether or not knowledge has limits. It is an institution separate from worries about truth or knowledge, in the same sense that the church is separate from worries about spirituality. These institutions exist because of their own social inertia. The scientific community was created as an institution to search for truths and it will continue to do so even when there are no new truths to discover, nor any problem to solve. If there were no problems to solve, both the problem and the solution would be created. If we know everything about what is within our ordinary experience, new expensive devices will have to be created to explore other phenomena beyond human experience. The rules that govern human groups do not depend on logical or rationalist arguments. The rules are as irrational as life itself. Why does a community of rodents grow and grow as long as food is provided? There is no reason for growing, it is just a rule of life. In the same way, the scientific community will grow so long as funding for research is provided, and will decrease when the funding is reduced. Each rodent can beget tens of new rodents. Scientists also beget tens of new scientists (their Ph.D. students), many of them will die scientifically (that is, they will not get a job as researchers); this is not dissimilar to the theory of evolution in living beings. The capacity to absorb all these new members of the community will only depend on how generous states are in feeding their scientists. This is nothing to do with the limits of knowledge but with the limits of power and persuasion. I do believe the institution of science will die too, but long after the decline of its spirit.

4.3 Other opinions on the limits of knowledge and the end of science

There is an extensive literature talking about the end of science, mainly in connection with the limits of knowledge. Indeed, many of the prophecies about this topic are more optimistic than Horgan's opinion or mine. Among the optimists, there is a broadly held view that science, far from being near its end, is constantly near the beginning of new interesting fields which will provide great moments of happy discovery in the future (e.g., Luminet, 2008). The physicist John Wheeler said, "As the island of our knowledge grows, so does the shore of our ignorance" (Horgan, 1996, ch. 3). This justifies a never-ending development of the dragon, for every time we cut off one head, several new heads grow up.

The pessimistic literature is also extensive. For instance, more than four decades ago, Stent (1969) contended that science was coming to an end; as well as technology, the arts, and all progressive, cumulative enterprises. When science seems most muscular, triumphant, potent, that may be when it is nearest death, said Stent, and I also think so. He stated that Linus Pauling brought chemistry to a state of completion in the '30s, and that, after the discovery of the structure of DNA, very few important things remained to be done in biology. Among his gloomier prophecies, he also announced that people would become more hedonistic, losing interest in science and the arts, an observation not very far from the trend we are witnessing nowadays.

In fact, this is by no means the earliest augury concerning the decline of science and culture. We will see in a later chapter that, in the 1920s, the thinker Oswald Spengler held a similar view, which influenced others who came after. Other authors think that belief in science is coming to an end, rather than science itself ending, as was the premise of a meeting in Syracuse, Minnesota, 1989 (Selve, ed., 1992) The philosopher Juan Arana sees the situation as follows:

> "The scientific community is beginning to become accustomed to the idea that, after many years of plenty, lean years are looming. Once the seas have become exhausted of easy truths - not discarding the opening up of other fisheries full of abundance — other arts and means must be sought of preparing longer hauls to trap more slippery fish". (Arana, 2012, ch. 8)

Most researchers actively working in some field complain about this pessimistic approach, and even ignore these opinions as eccentric. You can see, for instance, those particle physicists showing off their terribly expensive high-technology devices for experimentation, and proudly showing their results, such as the discovery of a new particle for the collection; this is a minor thing if we compare it with the development of quantum mechanics, for instance. They look like children with new toys. Per Bak, physicist, in an interview with Horgan (1996, ch. 8), said: "Most particle physicists think they're still doing science when they are really cleaning up the mess after the party". The physicist Lee Smolin said: "I am a member of the first generation of physicists educated since the standard model of particle physics was established. When I meet old friends from college and graduate school, we sometimes ask each other, 'What have we discovered that our generation can be proud of?' If we mean new fundamental discoveries, established by experiment and explained by theory—discoveries on the scale of those just mentioned—the answer, we have to admit, is 'Nothing'". (Smolin, 2006, "Introduction"). I agree. However, there are still people clinging to the illusion that science is nowadays producing great ideas, when what we are really gathering are secondary results, as a product of the application of highly developed technology to the great ideas of the past. The rest is propaganda. In other words, science lives on its own income, obtained a long time ago.

A more aggressive attack on present-day science comes from some authors who talk not only about the end of science but also about whether we should do anything to save it because of the way it has behaved in the last few decades (e.g., Oblomoff [collective], 2009; Ségalat, 2009). According to these authors, scientists have played the liberal and capitalist game and they are now victims of the system they have built. Research in science merged with industrial and technological advances, leaving theoretical work in a secondary position. This forced science to be pragmatic and opportunistic because science with technology depends strongly on economic support. The priority problems in science are those which obtain and move huge funds. This kills the creativity of ideas because research is oriented towards technological application in coalition with industry, following capitalist interests and general economic plans in which scientific research is just a part of a bigger system to create employ-ment in many factories and give benefits to certain industrialists. Research is converted into a race with commercial and sporting criteria: the most important thing is being competitive, to be first and to know how to sell a product. So scientists spend most of their time developing projects and evaluating the projects of their colleagues. These ideas of these authors are similar to those I have already expressed in Chapter 3. The authors conclude that it would be a miracle if good ideas emerged within this atmosphere, so they hope that science will reach the end soon, and claim that it is not worth making an effort to save it. In my opinion, science is still of great value and should not be confused with present-day scientific institu-tions, in the same sense that religion should not be confused with the church. I think we must not level the charge against science but against the capitalist system which currently supports it.

4.4   Knowledge as self-structure

Knowledge may be considered as an abstract entity which is beyond human brain activity. Of course, there is no knowledge without human beings (or other beings able to get knowledge), in the same sense that there is no life without organic stuff, or no humans with-out biological life, or not society without people. However, we may regard knowledge as an entity which has its own identity and which is self-organized: an autonomous structure. If we considered all the different supports of knowledge—books, computer files, etc.—we could even argue for it being entirely separate from human beings, but we must always bear in mind that human minds are necessary in order to read/interpret all that information. Philosophers usually carry out this kind of abstraction with new entities, which may appear as artefacts. Indeed, the consideration of "knowledge" as an entity is an artefact but so is the concept of "person", and many others which are useful in shaping our vision of the world. In fact, the only entity in nature is nature itself, in the form of matter-energy, but we can carry out an exercise in thought by isolating some particular structures and treating them as if they were autonomous entities, self-organized structures which make their own decisions.[1] There are many instances of this kind of abstraction: for instance, Dawkins' selfish gene (1976) or the Gaia hypothesis of Lovelock (1979) which posit the existence of entities (genes, Gaia) which govern life on earth. Here I do not pretend to claim that "knowledge" is an entity with its own will, with intelligence, a demigod controlling human beings for its own purposes—I am not a true metaphysicist—but it is helpful to consider the dynamics of knowledge, beyond the purposes of individuals.

There was a time when human beings controlled the science and knowledge they produced, and also controlled the other structures of society: production of goods, markets, etc. These were just tools for the needs of human society. Of course, there were some tyrants, some absolute authorities who had much more control than ordinary people, and there were some priests of knowledge who tried to control its flow, but in general one has the impression when reading about the history of civilization that human beings were behind the dynamics of the societies. Individual ambitions, will to power, conflicts of interests, treasons ... all these moved history. Nowadays, however, one has the impression that individuals are just simple marionettes whose strings are pulled by some abstract and superior entity. I am not talking about the Christian god but about that almighty god of modern capitalist times: money. Money is the great

---

[1] In my opinion, free will is an illusion, and all organisms or beings are simply fragments of nature, driven by its laws (López Corredoira, 2005). By "make their own decisions" I do not mean that these decisions stem from "free will", but I refer to feedback structures by which the internal and external conditions of the system are used to develop decisions in the behaviour of the system.

boss of our society. It governs the decisions of individuals and has much more power than the different nation-states or other human organizations. You may think that it is the richest people, those with the greatest amounts of money, who, in fact, control society, but those rich individuals are just as much cogs in the machinery of economics and prisoners of the system as poor people. Intuitively, one can see something like a "will of the capital", in which human beings are mere instruments in the execution on earth of such a will. Again, I insist, I am not talking about the emergence of a being in a metaphysical sense, but about the generation of structures in our society which have become automatic in some sense, and thus have become self-sustaining autonomous structures. Once implemented, they may work independently of the will of human individuals. The economy, our present financial system, appears in my opinion as the most important of the autonomous structures to have emerged in our civilization. A monster, Frankenstein's Creature, which once created, emancipates itself from its creators and turn against the interests of individuals, pushing them to slavery. We could live in a society in which individuals worked much less than they do now, but capital and its interest in unstoppable growth, pushes individuals to produce more and more, far beyond their needs. People ask for more and more work in a time when the work force is more dispensable than ever, given the high industrialization of our society; an absurd situation according to some thinkers, including myself (e.g., Krisis [collective], 1999; López Corredoira, 2009b). We are destroying our environment; our towns are becoming uglier and uglier, with so many factories, cars, concrete, etc.; our forest, seas, rivers, etc., are becoming contaminated; but we cannot stop the advance of the destruction because money pushes people to follow the career path of progress, technology and destruction, and few people want to be outside that system, few people want to live without money in their pockets.

It is not my place here to offer a dissertation about the economy, but I have mentioned this idea about money/capital because I think there is a certain parallel with the knowledge society. In my opinion, knowledge has become another autonomous structure within our society, although it is less important than money. In this new system of gods, money will be like Zeus, the major god, while the rest of the autonomous structures will be minor gods, depending always on the great god of money. Indeed, we must recognize that most of the people working as scientists in our society regard their research as a job, in order to earn a salary, and in most of the cases the obtaining of that salary and its associated status are the only things which keep them in their boring and passionless research fields. Knowledge emulates its major god, in the sense of becoming a continually growing structure, unstoppable, with no understanding about the real purpose of that growth. These social autonomous structures follow the rules of biological life, itself another autonomous structure, as with the example of rodents that I mentioned earlier, growing more and more with no more motivation than the growing itself. This is life. This is also the way that plagues propagate. Humanity may be considered as a plague if we think about overpopulation. And what is the purpose of having a world with a population of 7 billion persons instead of the 1.6 billion at the beginning of the twentieth century? Is humanity any better because of this? Probably not: the greater the amount of people, the greater the disturbance to one other and the lower the resources per capita.

Too much knowledge is also a problem. The times in which a simple man could control at least the most important aspects of knowledge have long since passed. Those research ideals have been left behind. Intellectual restlessness, the search for truth, created those colossi of knowledge who moved among the different fields like salmon in the rapids. Today, such moves have become impossible because knowledge has become heavy and sluggish. You will see an elephant sliding before you see a scientist being familiar with as many fields as our scientific forefathers were. Nowadays, a scientist has to specialize. Scientists have been specializing for quite a long time, but it is now a question of micro-specialization. The most a scientist can hope for is to achieve mastery of a few micro-specializations in which to invest their efforts or creative interests. It is hard to imagine anyone being drawn into a specialization because it is their only interest, unless the system has sent them crazy enough to believe that his topic is the centre of the world. This is clearly not so. Rather, it is more a case of converting the scientific process into an industrialized mass-production system. Everybody attends to his own cog so that the system runs smoothly. It is a betrayal of our scientific forefathers' ideals. Descartes gave science meaning to mankind as a search of truth in his *Rules for the direction of the mind*, and expressed in the first rule:

"Thus, if somebody seriously wants to research the truth of things, he must not choose an unusual branch of science, since all of those involved are known to one another; rather, he must think only about increasing the natural light of reason, not to solve this or that school difficulty but to get an understanding about life that shows us the behaviour we have to choose".

That is, science has abandoned wisdom and become a mere technical profession. It is supposed that humanity has become wiser because of the greater amount of knowledge gained, but it is not so. In general, scientists and philosophers in the past were much wiser, even with less knowledge. Nowadays, we have many Ph.Ds in science, and statistically we might suppose that we live in a wiser society because of that. However, a more careful consideration of the question would show that most of these scientists are simple workers without that much culture.

For a single individual, scientist or not, the increase in knowledge is not a benefit but a problem: they may get lost among the huge quantities of information. We will find a lot of rubbish before we find an intelligent new idea among recent scientific results. But the amount of information continues to grow anyway. As mentioned in Chapter 3, according to some philosophers of science, every fifteen years approximately, the mass of scientific knowledge is doubled (Iradier, 2009), which means than in a century the production of scientific results is multiplied by a factor around one hundred. One may conclude that for individuals the benefit is possibly not clear but for humanity as a whole it is worthwhile, because at least we know there is somebody on earth who find a benefit in every single piece of knowledge. This thought is precisely what gives rise to the entity of knowledge as a self-structure. At the moment in which we think about the general purposes of a society (the growth of GDP, the growth of knowledge, etc.), without thinking about the non-financial benefits to ourselves, we are delivering our souls to abstraction, we are becoming soldiers of an army which does not defend our particular interests. And this raises a question about the army: what is the point of fighting in a war which is not our war? Are we just mercenaries in a professional army, because "they pay", or do we fight with passion to defend our ideas or our land? Something similar may be said about scientific research.

The philosophy implicit in the name of the "knowledge society" is simply a belief, a prejudice without too many arguments in its favour. Apart from technological progress—which, as I have said, I do not consider to be generally positive; I do not consider twentieth-century technology has improved the world— there are no valuable returns for the collective wisdom of the nations. The pursuit of ever more knowledge of smaller and smaller details in a scientific field is more an obsession and social inertia than it is an intelligent plan of enlightenment.

Some authors (e.g. Llinás, 2001) have compared the internet with a neurological system in the transmission of information. The internet may be another autonomous structure which takes its own decisions. Llinás (2001) says that the internet and other mass media contribute to the homogeneity of thought, and to a hedonist society, a decadent and sybaritic civilization which may fall toward self-destruction and forgetfulness, something similar to what happens when a laboratory mouse is connected with electrodes which stimulate the pleasure areas of its brain when it presses a button: the mouse obsessively presses the button until its death. What a pessimistic prophecy! But this is indeed the kind of thing that happens with obsessions. A society which is continuously obsessed with getting money, data, knowledge, and forgets the purpose of life, will not have a much better end than the mouse. The internet is indeed a subgroup of the contents of our present-day knowledge/information society. Everything can be found on the internet, everything spread in millions of sites, in huge disorder and bias, everything except a unified worldview which gives sense to our lives.

Scientific knowledge is something good; I think I have expressed this idea with sufficient clarity in Chapter 2. But anything in excess is bad. Water is good for living beings, but an excess of water drowns them. Therefore, my message here is different from Horgan's main one: Science is not being eroded by the limits of knowledge but by unlimited knowledge. Certainly, there is a limit to the number of really interesting questions that can be answered, a limit which we have already surpassed, but society will not stop its activity just because of that limit; rather, I think society will stop the obsessive collection of knowledge when it feels lost among the information, when it gets fed up with getting drunk after so many science parties, when it falls down exhausted and wonders whether we are servants of the structures we have created or we control them, whether man was made for science or science made for man, as Unamuno said.

5 ORTHODOX AND HETERODOX SCIENCE

Science is not a direct means for reaching the truth. Science works with hypotheses rather than with truths. This fact, although recognized, is usually forgotten. It gives rise to the creation of certain key groups within science which think that their hypotheses are indubitably solid truths, and think that the hypotheses of other minority groups are just extravagant or crackpot ideas. These are usually referred to as the orthodox and heterodox positions in a given field. Certainly, there are fields in which, after a period of development, things become very clear and hypotheses are converted into solid theses. For instance, atomic theory or the bacteriological origin of some illnesses are nowadays indubitably true. But all through history, and even now, there have been many instances of discussion about how to interpret aspects of nature, with various possible options without a clear answer, in which a group of scientists have opted to claim that their position is the good or orthodox one while other positions are heresies. Again, we find another behaviour common to scientific and religious institutions.

A lack of historical perspective may have originated the belief that science advances in a straight line; that important theories, once they are accepted by most of the community, are a part of knowledge forever. Clearly, this is not the case. False theories may last for centuries or even more than a thousand years and still be accepted without any doubts. In Chapter 2, I gave some examples of this, Ptolemaic geocentric astronomy being one example. There have also been many other theories which were important for a long time, but which were eventually demonstrated to be wrong: the phlogiston theory, the caloric theory, Newtonian optics, the proposal of the existence of an "ether", etc. The opposite thing is also true. There are ideas which have been rejected, and forgotten for a long time, until they are later recovered and become successful ways of explaining phenomena, examples being the heliocentric theory, tectonic plate theory, etc.

There is much to learn about contemporary science from examining the roles of orthodox and heterodox scientists nowadays. Apart from purely scientific considerations, from a sociological/anthropological point of view the phenomenon is a vivid expression of a fight to acquire or preserve power or status. Some dominant theories serve the interests of powerful groups, helping to legitimise business as usual. All hierarchical social structures show this feature, and the higher the level of institutionalization, the lower the level of freedom for individuals to undertake research away from the mainstream. The maximum amount of control of science is associated with the minimum amount of space for creativity and new ideas. Certainly, we are used to listening to news about innovation and new discoveries in science, but these are mostly technical applications of old ideas, or developments within the old rules of the same game.

5.1 Have you made a revolutionary discovery? Come in! Come in! We were waiting for you...

There are many naïve persons, scientists or non-scientists, around the world who still believe that science is an open process in which the best ideas are quickly recognized and accepted, while the wrong ideas are immediately discarded. This kind of individuals thinks that achievement in science nowadays depends on intelligence, on genius. They think that someone could be working hard in a laboratory, or developing some theoretical idea and, if they were to a revolutionary discovery, they would open the door of the room in which they was cloistered and shout along the corridors "Eureka! Eureka!"; then, colleagues would approach and say: "Have you made a revolutionary discovery? Come in! Come in! We were waiting for you ..." and the genius would have the chance to show their new discoveries and their colleagues would open their mouths, surprised by the new idea, recognizing its merit, and carrying the genius on their shoulders while shouting "Torero, torero,...!" ("Bullfighter" in Spanish).

This was never the way in which general ideas were accepted, and nowadays it is even further away from reality than ever. Certainly, all through history, there were revolutionary ideas which were eventually accepted by the scientific community, but rarely within a very short period of time. Generally, some generations have to pass before a shift in important ideas is produced. The process is slow, and this slowness is related to the psychological structure of individuals and the associated sociological structures of the scientific institutions. There is an inherent inertia in human brains about accepted world views which take a long time to be changed. You cannot wait, as Galileo had to, for the old authorities, who had spent decades studying Aristotelian physics and astronomy, to suddenly change their opinions about the world, even if they were shown the satellites moving around Jupiter or the craters of the Moon with the telescope. "Faith is the organ of knowledge, intellect is secondary. Your science without premise is a myth", said Naphta, a character in Thomas Mann's novel, *The Magic Mountain* .

It seems that there is a historical mechanism that, as time goes by, and independently of human interests, filters out and polishes the most reliable knowledge. Probably, it is because vested interests gradually vanish with the advance of generations, and it is only after some decades, or perhaps even hundreds of years, that ideas with intrinsic value are distilled and survive to give us their wisdom. Indeed, history is not always fair. Many good ideas are forgotten and are not recovered until they again rise to the surface of independent thought. There are also many cases of historically famous researchers who have stolen credit from unknown people. Neither is history itself perfect for, after all, it is also human.

In the present day, we have an important advantage: There is no direct censorship, and nobody is burnt at the stake because of their ideas. However, there is significant indirect censorship; for instance, through rejecting the publication of interesting heterodox ideas in mainstream journals. Even worse, the huge volume of knowledge does not allow us to see the information which is not supported by propaganda, in most cases getting lost the space of a generation.

Other forces are working against the creation of new ideas in science. The machinery of the system is becoming more and more collective, moving away from individual scientific adventures. Each scientist is more like a cog in a big system than a free intellectual. Most scientists are busy with administrative and routine tasks within a predetermined project. They do not have much time to explore new ideas, and if they have decided to explore new ideas and they find something of interest (which will be quite rare, because most of the interesting ideas have already been explored; although vanity may make these researchers believe in the importance of their results), the difficulty comes in finding other colleagues to read their work and analyze the plausibility of new ideas.

On the one hand, heterodox scientists are possessed by a feeling of being an unappreciated genius, have too much "ego", normally working alone/individually or in very small groups, creative, intelligent, non-conformist. A vast majority of them are men. Their dream is to create a new paradigm in science, something which completely changes our view of the field of science in which they are working.

For instance, there are many of them who are trying to show that Einstein was wrong, maybe because he is the symbol of genius and defeating his theory would mean that they are greater than Einstein. Most of them are crackpots. On the other hand, orthodox scientists, who constitute the majority of the community, are dominated by groupthink[1] and snowball effect, following a leader's opinion as in the story of the emperor's new clothes, are good workers performing monotonous tasks without ideas of their own in large groups, are specialists in a small field which they know very well, conformist, domestic. Their dream is to get a permanent position at a university or research centre, to be the leader of a project, to be a recognized science administrator. Most of them are like sheep, some of them with the vocation of shepherds as well. Luminet (2008) compares these people doing "normal science" to craftsmen, and compares those scientists who pursue a revolutionary science to imaginative artists. The sociological reasons for favouring mainstream to orthodox craftsmen might also be related to the preference for domesticity in our civilization. Sheep rather than crackpots are preferred. Finding a promising change of paradigm which will bring us closer to the truth among thousands of crazy proposals is very difficult; in ortho-

---

[1] In a sociological analysis, Janis (1972) categorizes the symptoms of groupthink as: 1) An illusion of invulnerability, shared by most or all the members, which creates excessive optimum and encourages the taking of extreme risks. 2) An unquestioned belief in the group's inherent morality, allowing the members to ignore the ethical or moral consequences of their decisions. 3) Collective efforts at rationalization in order to discount warnings or other information that might lead the members to reconsider their assumptions before they recommit themselves to their past policy decisions. 4) Stereotyped views of enemy leaders as too deviant to warrant genuine attempts to negotiate, or as too weak and stupid to counter risky attempts made at defeating their purposes. 5) Self-censorship of deviations from the apparent group consensus, reflecting each member's inclination to minimize to himself the importance of his doubts and counterarguments. 6) A shared illusion of unanimity concerning judgments conforming to the majority view (partly resulting from self-censorship of deviations, augmented by the false assumption that silence means consent). 7) Direct pressure on any member who expresses strong arguments against any of the group's stereotypes, illusions, or commitments, making clear that this type of dissent is contrary to what is expected of all loyal members. 8) The emergence of self-appointed mindguards—members who protect the group from adverse information that might shatter their shared complacency about the effectiveness and morality of their decisions (Dolsenhe, 2011, ch. 12).

dox science, although absolute truth is not guaranteed, at least a consensus version of the truth can be offered.

The reality is that nobody is waiting for revolutionary ideas, they are not welcome, now less than ever, and the difficulties that professional researchers have when they want to challenge dominant ideas (e.g., Campanario & Martin, 2004) are enough to dissuade them in their enterprise or cause to be rejected as outsiders by the system.

5.2   Amateurs in science

A significant number of professional researchers are caught up in fights to defend heterodox theories, although the ratio in comparison with orthodox scientist is low because of the reasons given above. However, there is, in all polemical questions of science, an attraction for amateurs who want to play the game of science. Usually they try to mask their ignorance and reject the criticism of them by presenting themselves as unappreciated geniuses, with a capacity for thought beyond that of most scientists. The problem for amateurs in the sciences is the vanishingly small number of possibilities available to observe or conduct experiments, not to mention the bad reputation associated with free-thinking that occurs away from official institutions. Since the expenses for the necessary materials are very high, the possibility of doing high-level empirical research on the margins, in any field, is practically zero. The only possibility is pure theory/speculation, or perhaps working on empirical data produced by other scientists, which in fact happens quite frequently.

I often receive, by email, new theories from amateurs who want to topple all the well-established tenets of physics to leave space for new and often ridiculous theories, or else cases of cosmological theories that fail in even the most basic aspects. This kind of work makes almost no reference to professional journals, and tends to cite popular books on science. Rather than studying a particular scientific problem, they talk about very general matters. Precisely because of that, independent research carried out away from official institutions finds problems with credibility; for every well-prepared researcher who wants to do serious things, there are thousands of "barmies" on the planet who dream of creating a theory of physics inspired by the heavens, which demolishes the past and opens a new era in the history of science. I once heard an interview on a radio program with a carpenter who had never studied physics, but had just read some popular books on physics, without trying to understand anything about mathematics. The carpenter said that he had written seven books about black holes, and he complained that he could not publish any one of them. I do not want to judge negatively the efforts of amateurs, who perhaps have read a popular-science book by Hawking and think that they are able to work as researchers. I do not want to be a part of a system that castrates any attempt at originality just because it is challenging. Nevertheless, the reality is that amateurs' theories have a lot of weaknesses and inconsistencies because they do have not enough knowledge of the subject. And the result is that the thousands of "barmies" in the shadows do not get to hear the voice of some possible genius who might be in their midst. As it is, autonomous research activities are not taken seriously, and one must use official mechanisms in order to be heard by other professional researchers.

Nonetheless, I appreciate the spirit of most amateurs. They behave like the scientists of the past. And when they are able to mount a congress, paid for from their own pockets, one has the impression that a new Solvay Conference is taking place, given the number of revolutionary new ideas which are shown. They still feel the enthusiasm of scientific creation. It is a pity that this enthusiasm is not present to the same degree among professional scientists. And it is a pity that almost all the revolutionary ideas are at first sight crazy ideas, and that the few of them which do not seem to be crazy ideas at first sight are most likely crazy ideas when one studies them in detail. It is a pity that almost all these amateurs or semi-amateurs (some of them have some formal education in scientific disciplines but are not specialized researchers) live in a fantasy world, rather like Don Quixote in their enterprises. Certainly, Quixote's character is very noble and his idealism admirable, but he does not live in the real world. In the same sense, amateurs do not live the same science as that which is followed by professional researchers.

5.3   Conservatism and consensus science

In general, science is a conservative system. It is a system based on tradition, on cumulative knowledge, on the patient accumulation of wisdom which filters out the bad information and distils the good. Science does not work like the arts, in which spontaneity and novelty are more important than tradition. Science is not philosophy, in which pluralism is accepted and in which tens or hundreds of ideologies live alongside one another. Science is not plural, and that is precisely its advantage over philosophy. It pursues the truth, and there is only room for one truth. This allows for some advance rather than science becoming stagnant, caught in an eternal discussion like philosophy. There is a unity in the interpretation of nature and this allows us to use the same laws of physics, the same chemistry, etc., to pursue fresh knowledge; and it works. If we lived in a chaotic world without any regularity maybe an artistic approach would be more appropriate, but since we live in a universe with regular laws which can be deduced empirically or from an appropriate theoretical frame, it is possible to find the rules of observable order in the cosmos.[2] As noted by Horgan (1996), science poses questions and resolves them in a way that critics, philosophers, historians cannot; discussion in literature, for instance, about the meaning or interpretation of a text can never be resolved. This is good, and it is one of the keys to the success of science.

"Science's success stems in large part from its conservatism (...) The scientific culture was once much smaller and therefore more susceptible to rapid change. Now it has become a vast, intellectual, social, and political bureaucracy, with inertia to match". (Horgan, 1996, ch. 5)

Conservatism and tradition in general are also good for society. Our culture is a cumulative process in which we draw profit from our civilization's past. We cannot and should not destroy all forms of society and begin again from nothing. It is important to keep most of what we have learned from the past, the works of the past, our languages, and so on. In the political/economical order, conservatism has also an important role, that of stability, of equilibrium.

Nonetheless, some degree of progressiveness is also necessary. A pure conservative position, without any progress, would keep our society stagnant, and all of us know what happens to stagnant water: it tends to form bogs. Hegel used this metaphor to indicate that a society needs, from time to time, to rise in revolt against the establishment. Keeping one form of politics for a long time accords too much power to certain sectors of society, and it leads to corruption of the values which originated that form of politics. No form of

---

[2] In its most general sense, a "cosmos" is an orderly or harmonious system. It originates from a Greek term "*κόσμος*" meaning "ordered world" and is the antithetical concept of chaos.

politics, not monarchy nor democracy + capitalism nor communism, etc., is good forever. Humanity must, from time to time, invent new forms, and social revolutions and wars are unavoidable in this process.

With science we could make similar observations. Although, certainly, there is an important difference in relation to politics: In politics, there are no absolute truths; in nature, there are absolute truths and science tries to understand them. Certainly, there are some people who try to convince the rest of the humanity that our present-day form of politics (democracy + capitalism) is universal and the only way to establish a fair order in our society, and that the Universal Declaration of Human Rights is an absolute statement as to what human beings should be; but these are simple prejudices based on ideology, not truths in the positive sense that science has achieved. Science has achieved some truths which will possibly last forever, and no revolution will be able to deny them. Atomic theory will last forever, I guess; I think Archimedes' principle of hydrostatics in incompressible fluids will persist forever; the bacteriological origin of some illnesses will be certain forever; etc. This explains why science is a more conservative system than other human affairs. Only in some fields where the truth is not so immediately clear—for instance, the origin of life, cosmology, the final components of matter, etc.—is there the possibility for some kind of revolution, and this is the place where conservative scientists try to defend their power against the possible revolutions of progressive ideas.

It is claimed that our theories are very robust, with the intention of transmitting an impression of security in our knowledge. People are usually afraid of uncertainties; they want secure information, a secure world. They prefer to live in a world without surprises, with a guaranteed retirement salary, a world in which one can live a long life surrounded by children and grandchildren, and die quietly in a bed after that long life. And conservative science offers this view of calmness, in which everything is apparently under control, and the few discordant voices are just those of madmen. They claim that standard theories are robust because of their predictive capacity and unity. However, in my opinion, there are reasons to become rather more nervous of this calm orthodoxy. Many standard theories are being built *ad hoc*, and their predictive capacity is only a consequence of the increasing number of free parameters. For instance, the standard particle model has around twenty parameters; it has lost its simplicity and predictive capacity and unity (Smolin, 2006; Iradier,

2009; Unzicker, 2010, ch. 12). The origin of many of these parameters remains unknown and, while no significant evidence for a failure of the theory has emerged so far, it may be incomplete because it does not account for facts such as the oscillations of the neutrino; it accounts for the electromagnetic, weak and strong interactions, but fails to unify this last with the two former and is also understood to be incomplete because it does not account for gravitation or dark energy; neither does it contain any viable dark matter particle (Yao, *et al.*, 2006). Something similar happens with cosmology (López Corredoira, 2009a). In total, it is stated that particle physics and cosmology have between them more than thirty independent fundamental constants (Tegmark, *et al.*, 2006).

Sometimes, there are even clear cases of disagreement with the observations/experiments which are arranged *a posteriori* with the introduction of a modification in the theory. For instance, the Weinberg angle or weak mixing angle was a putative constant in the theory of the electroweak interaction which gives the relationship between the masses of the W and Z bosons. However, the Weinberg angle was measured with different values for different experiments and with respect to the predictions, and this was justified by saying that the "constant" changes with the energy of the experimental measurement (Unzicker, 2010, ch. 12).

Nature is difficult to understand, and the truth may have nothing to do with the positions taken by some scientists who do not want to have disturbed the peaceful tranquillity of their lives. Simplicity and elegance are not sufficiently good arguments. Ockham's razor is usually cited by those who pretend to emulate a philosopher. Apparently, Ockham and his razor is the only philosophical reference that many scientists can lay claim to but it is often cited inappropriately because nature's simplicity is confused with the simplicity of what they are calculating. "Nature does not care for analytical difficulties", said Fresnel in 1826.

Who decides what is or is not an absolute truth in science? That is the key question. On the one hand, conservative sectors which have the power to control science will always claim that their ideas are undeniably absolute truths; on the other hand, the few voices of the revolutionary creators of new ideas, mixed with the many voices of crackpots, will cast doubt on anything claimed as absolute truth. Think, for instance, about extreme ideas, such as creationism, which claims that the earth is only six thousand years old and denies the evolution of the species. The fact is that there is no absolute criterion for separating absolute truths from false hypotheses, although there are some quite clear cases.

In spite of the general conservative approach, science has produced some revolutions. Now, in the wake of its institutionalization, science has become more conservative. The consensus science of big groups of scientists has occupied the privileged position, pushing aside the individual ideas of autonomous scientists. The snowball effect leads to a science of consensus in which the important thing is convincing the mass of scientists through mass media sources dedicated to them: reviews, conferences, etc. And, as pointed out by Luminet (2008), a consensus in the scientific community is transformed into orthodoxy and ends up becoming counter-revolutionary.

This counter-revolution is constituted as a mechanism of self-defence created by the autonomous structure of knowledge in the building of theories. According to Iradier (2009), all disciplines or specialisations were constituted and directed by a process of removing questions. Once a theory is established by the system, its fundamentals are not investigated any further. The geological stratification of theories is a great obstacle, says Iradier (2009): While a theory of science as an exploration without restrictions or prejudices and as a progressive approach to the truth is given publicity, research on the fundamentals of science is despised, and tricks are played against the people who have alternative ideas to these fundamentals. The first thing to be established becomes the last thing that is allowed to be revised.

Curiously, the most conservative people are not usually the oldest, but the youngest generations. For instance in cosmology, older scientists (more than sixty years old) are much more likely to discuss a crazy idea about the universe than anyone under the age of forty (Hogg, 2009; Loeb, 2010). Young astrophysicists nowadays invest almost all of their research time conservatively, in mainstream ideas that have already been explored extensively in the literature, and they do not dare, or only very rarely attempt, to explore new ideas (Loeb, 2010). Although the same phenomenon existed in previous decades, it is becoming alarmingly prevalent today because a growing percentage of observational and theoretical projects are pursued through large groups with rigid research agendas, added to which is peer pressure and job market prospects. This sociological phenomenon arises from the fear of younger generations that they might jeopardize their careers or risk difficulties to gaining a grant or position if they are involved in heterodox ideas; whereas older generations do not have to worry about their jobs and can take greater risks. It is also the fact that young people are usually more naïve; as remarked by Schopenhauer (1851) to the know-alls of his time: "Every thirty years, a new generation of sincere, chatty people, ignorant of everything, wants to devour hastily and summarily the results of human knowledge accumulated over centuries, and immediately they think themselves more skilful than everyone in the past". Also, older people have greater respectability and people do not play tricks against them when they suggest a crazy idea. But this fact also may reflect that conservatism has advanced very fast in the last few decades, pushing the science of the twenty-first century into a much more conservative politics than that of the twentieth century.

In particle physics, something similar or even more extreme has occurred. According to Schroer (2011), at the time of Pauli, Landau and other important physicists, a flawed idea about particle theory could not have survived for long in that critical environment but, since the 1980s, personal disputes about particle theory became rare and as a consequence the fundamental critical feedback has slowly disappeared. The only remaining criterion for the quality of a theory, says Schroer (2011), is the reputation of its proponents and the growth of their community, meaning that theories originating outside the monocultures of big communities do not receive any attention.

This is unavoidable. The evolution of science towards a tyranny of knowledge is unavoidable, because of the logic of the construction of knowledge itself, both because the knowledge is becoming more and more solid, and also because the dominant groups within the dominant ideas in science are becoming more and more powerful. The distinction between power and truth in science is difficult. Anyway, it is not good, because some level of progressiveness and revolution is necessary to keep science alive. To do research is to fight; what else do human beings do? To fight against the power of others or to attain our own power, that depends on us. Science can be a revolution or deadlocked in idleness. Still waters tend to form bogs (López Corredoira, 2008b).

5.4   Critical mass in a self-structure of knowledge

The situation today is as follows. On the one hand, knowledge is produced more quickly than ever. On the other hand, the psychological mechanisms which accept new world views work neither quicker nor slower than in the past. We have an increasing flood of information which needs to pass through the narrow hole of our mind, which has not increased. What is the result of this situation? The situation is that society is drowned in ideas and information without assimilating any of it, and only a few ideas, those selected by the establishment, will make some impact. We are in the era of mass media and propaganda. Only science which is given publicity in newspapers, TV, etc. will have any resonance in our society. But in order to control the mass media one needs money and social status. Only administrators and science politicians, who worry more about getting funds and prestige than about solving scientific problems, will be able to pass through this filter. People who dedicate great effort and time to thinking about new ideas in science are not paid enough attention. It is not the time of individual scientists. It is the time of big corporations, of megaprojects which know how to make the maximum profit from state funds.

Among the tons of written papers filled with scientific results, only a few will have sufficient impact to attract the interest of society and maintain continuity. Something similar to the theory in evolution in animals can be applied to knowledge. Also, we can compare the development of ideas with nuclear chain reactions, in which a critical mass is necessary for a reaction to take place. The critical mass is the smallest amount of fissile material needed for a sustained nuclear chain reaction. An equivalent, in the kingdom of knowledge would be the minimum number of people required to endorse an idea in order to make some kind of impact with it. If the group of people is very small, sometimes only one individual, what most commonly happens is that these ideas are not given enough attention and never reach the main body of the scientific community, being diluted in a short time. When the group of people promoting the idea is large enough, the transmission of information is more efficient. Scientists will tell other scientists or they will publicize results at conferences, ensuring that the news reaches the main body of the scientific community on several fronts. It is the moment at which the topic becomes widely discussed, when conferences are organized around that topic, and the impact continues growing unless it is stopped by a counter-argument from another group.

It is true that certain standard theories work better at explaining the data. However, the number of persons involved in working on a particular theory, and in patching it here and there is not generally known. Standard theories are supported by thousands or tens of thousands of researchers who in some way are involved or interested in the theory being correct, otherwise their life's work would be jeopardized. However, an opposing theory may be defended by a small number of researchers. Even if they are very good scientists, they cannot compete to produce patches and to spread propaganda on the same scale as the huge numbers of orthodox researchers. These researchers have to fight against the system without money, without students, without access to expensive devices for observations and experiments.

The effect is very similar in some respects to the cultural industry, for instance, in the production of books. Big publishers understand very well the mechanics of producing best-sellers and successful books with high impact, even if their content is rubbish; although, usually, these books die in a period of time inversely proportional to the time in which success was achieved. The mechanism is simple: rapid publicity bombing of the book through many mass media outlets, interviews with the author; printing many copies of the book and getting it exposure on the bookshelves of as many bookshops as possible; the chain reaction is quickly obtained after the initial explosion and, later, propagates simply through the recommendations of readers. Commonly, people begin to talk about the book. And when two or more friends talk to you about a book, you have the impression that: "Ah! This is an important book of our time, because everybody is talking about it", and you buy the book, read it, and talk about it to other friends. Possibly, you will realize that the book is rubbish, but the important thing is that you talk about it, and thus you contribute to its diffusion. On the other hand, somebody could have written a masterpiece, but if it does not reach the critical mass and get the social reaction, it will be quickly forgotten. There are exceptions to these rules, but they are very few.

The same thing happens in the production of science for the market of scientists. Big groups making mediocre science know that the key to getting two weeks of fame is a quick bombing of information through the community in order to achieve the critical mass necessary for a social reaction. They make tours like *supervedettes* and they do get fame for a period of time, although this is likely to be quickly extinguished.

The effect of this mechanism of social reaction achieved by critical mass is that orthodoxies, with more money and more people around them, get stronger, leaving little place for alternative views. Only science made with a lot of noise is visible; the rest is neglected. This includes the "Gold" effect (Kundt, 2008), by which a simple unqualified belief can occasionally be converted into a generally accepted scientific theory through the screening action of refereed literature, meetings planned by scientific organizing committees, and through the distribution of funds controlled by "club opinions".

This also includes cases in which competing heterodox scientific theories are attacked by supporters of dominant scientific theories, which have sufficient critical mass to be listened to, while the responses of heterodox scientists are not heard because they cannot reach that critical mass. To reduce concern about their actions, and avoid to be seen as unfairly abusing the heterodox scientists, supporters the dominant theory can use a variety of techniques (Martin, 2010): conceal the violation of expectations; devalue the competing theory and its advocates; represent the process as their own; use expert panels, meetings and other formal processes to give a stamp of approval to the dominant view; and intimidate opponents.

Another obvious effect is the endless appearances of bogus bright stars. Scientists and ordinary people notice these frequent appearances, and they observe later how the importance of the discoveries announced with a lot of fuss is much less than implied by the publicity. After the sparkling moment of euphoria vanishes, there remains the bitter taste of emptiness. There are still some naïve people who think that we are in a golden age of science to judge by the number of big discoveries announced, but with time the number of such people decreases and there remains an impression of saturation of nothing, saturation of petty science (as well as lightweight literature and other products of the culture industry). This is a consequence of the predominance of the "critical mass" rule in the transmission of knowledge. The usual thing in a situation where there is only a small amount of information would be that important discoveries were propagated slowly, with a good deal of difficulty, only reaching the main body of the community after a long time. But this mechanism was perverted, because the slow propagation of information leads to extinction when surrounded by an overflow of information, and now the mechanics of the mass media is predominant in the information war. Scientists tend to exaggerate the importance of their discoveries because it is the only way to survive in the information jungle.

5.5 The real method of science

Basically, there are two different methodologies in doing physics, both inherited from different ways of thinking in ancient Greece:

- The mathematical deductive method: this is the method employed by Pythagoras and Plato. The pure relations of numbers in arithmetic and geometry are the immutable reality behind changing appearances in the world of the senses. We cannot reach the truth through observation with the senses, but only through pure reason, which may investigate the abstract mathematical forms that govern the world. An example within modern science might be Einstein's theory of general relativity, which was posited from aesthetic and/or rational principles at a time in which observational data did not require a new theory of gravity.

- The empirical inductive method: this is the method employed, for instance, by Anaxagoras in understanding nature. Aristotle uses both inductive and deductive methods, and he says that "the mathematical method is not the method of the physicists, because Nature, perhaps all, involves matter" (*Metaphysics*, book II). Matter is not numbers or mathematics. Nature should be known through observations and extrapolations made from them. The empiricism of Galileo Galilei might be an example within modern science, although all scientists, even Galileo, are also partly Pythagoreans.

In principle, most physicists recognize the empirical method as the best one. However, in practice, there are many cases of camouflaged Pythagoreanism. As Sherlock Holmes, the fictitious character of the novels by Sir Arthur Conan Doyle, said: "Before one has data, one begins to twist facts to suit theories instead of theories to suit data" ("A Scandal in Bohemia"), and "It is a capital mistake to theorize before you have all the evidence" ("A Study in Scarlet"). It is a capital mistake but people make it, nonetheless. And what happens in physics is also extrapolatable to other sciences, with the peculiarity that instead of mathematics other tools are used, but with the same scheme of non-objective prejudices from which theories are deduced rather than a free interpretation of data in an inductive method. It is well known by many sociologists of science or the scientists themselves (e.g., Mulkay, 1976) that science does not follow scientific methods nor the premises of objectivity. There is plenty of evidence of specific biases, for example a bias in support of papers confirming the currently accepted viewpoint or in favour of established researchers (Armstrong, 1997). Henry Bauer (2012) thinks that the nature of scientific activity has changed dramatically over the last half century, and the objectivity and rigorous search for evidence that once defined it are being abandoned, in a scientific environment where distinguished experts who hold contrary views are shunned. According to Bauer (2012), in some areas, such as Big Bang cosmology, human-caused global warming, HIV as a cause of AIDS, and the efficacy of anti-depressant drugs, dogma has taken the place of authentic science, and conflicts of interest have become pervasive in the world of science. Another example: the dominant medical view about cancer legitimises conventional treatments, using surgery, radiotherapy and chemotherapy, and thereby serves the interests of the medical profession and related industries (Martin, 2010).

According to Iradier (2009), prejudices of contemporary scientists include: 1) only by constructing experiments of great sophistication can new knowledge be obtained; 2) actual equations are formally excellent and desirable; 3) science sieves subjective elements to obtain only objective truths. The first prejudice is a reflection of the interests of the technological industry seeking to profit from funds dedicated to research in pure science, and it assumes that all science which can be done with cheap experiments has already been done and is correct. The second prejudice also emphasizes the assumption of the correctness of the basic rules which are in foundations of physics or other sciences. The third prejudice is based on a very idealistic idea of an unbiased science, far from social reality.

It is usually claimed that science is objective and, therefore, one theory is supported rather than another theory because further proof has been obtained in favour of the first one rather than the second one. These words sound very nice, and they even appear honest. However, one must consider that not everything is so honest, but neither do I say that everything is pure manipulation. No, there are many cases in which nature shows itself clearly enough in experiments and observations, and the conclusions are irrefutable. But there are many muddy cases, in which the power of manipulation is stronger than nature. In practice, the method to be applied is usually not very objective and basically is as follows (López Corredoira, 2008a):

• Given a theory A self-proclaimed orthodox or standard, and a non-orthodox or non-standard theory B. If the observations/experiments achieve what was predicted by theory A and not by theory B, this implies significant success for theory A, something which must be divulged immediately to the all-important mass media. This means that there are no doubts that theory A is the right one. Theory B is wrong; one must forget this theory and, therefore, any further research directed to it must be blocked (putting obstacles in the way of publication, giving no access to instruments, no more funds for it, etc.).

• If the observations/experiments achieve what was predicted by theory B rather than by theory A, this means nothing. Science is very complex and before taking a position we must think further about the matter and make further tests. It is probable that the observer of such had a failure at some point; further experiments or observations are needed (and it will be difficult to make further tests because we are not going to allow the use of expensive devices to retest such a stupid theory as theory B). Who knows! Perhaps the observed thing is due to effect "So-and-so", of course; perhaps they have not corrected the data from this effect, about which we know nothing. Everything is so complex. We must be sure before we can say something about which theory is correct. Furthermore, by adding some new elements to theory A surely it will also predict these observations, and, since we have an army of theoreticians ready to put in patches and discover new effects, in less than three months we will have a new theory A (albeit with some changes) which will agree the data. In any case, while we are in troubled waters, and as long as we do not clarify the question, theory A remains. Perhaps, as was suggested by the heterodox cosmologist Halton Arp (2008), the informal saying "to make extraordinary changes one requires extraordinary evidence" really means "to make personally disadvantageous changes no evidence is extraordinary enough".

6   THE DECLINE OF SCIENCE

The last three chapters have shown some of the peculiarities of science nowadays. We can summarize the present-day situation described in these chapters with the following three statements:

1. Society is drowned in huge amounts of knowledge, most of it being about things of little importance for our cosmic vision, or producing no advances in the basic fundamentals of pure science, only technical applications or secondary details.

2. In the few fields where some important aspects of unsolved questions have arisen, powerful groups control the flow of information and push toward consensus truths rather than having objective discussions within a scientific methodology; it gives few guarantees that we are obtaining solid new truths about nature.

3. Individual creativity is condemned to disappear in favour of big corporations of administrators and politicians of science specialized in searching ways to get money from States in megaprojects with increasing costs and diminishing returns.

We can use one adjective to describe the status of science at present and in the near-future: decadent. It is only a subjective perception. Possibly other people will think the opposite thing, that we live in a golden age of science. Rather than a question of argument, it is a question of sensitivity, of being able to perceive the sense or non-sense of the major enterprises which are nowadays called science from a human point of view. The quantity of publications, the quantity of big instruments and the technology created, the number of jobs created in research, the accurate control of our science in comparison with past times, etc. might be arguments to show that science is presently living in a wonderful epoch. However, I would reply, the spirit of science is being lost. And how do you measure the quantity of spirit? No, it is not a measurable quantity; forget about creating a new scientometric method to determine the amount of scientific spirit. It is a question of sensitivity: just look around; just talk with some leading scientists and observe their lives, their work. Technocracy is replacing the joy of scientific creativity.

The same thing could be said about poetry: Do you think we live in a golden age of poetry now because a huge number of poems can be found on the internet, there are a lot of poetry competitions with hundreds of participants, and there are many poetry clubs, etc.? No, the spirit of poetry is nothing to do with that. What then is? If you cannot find an answer yourself, it is because you are not sensitive enough to poetry. Something similar happens with sciences. It is necessary to be sensitive to scientific thought in order to appreciate its boom or decadence. In this chapter, I want to make further observations and reflections on what might be called decadence, decay or the decline of science.

6.1   The triumph of bureaucracy and mediocracy in Science

Capitalism is all-consuming, and its enemies have two destinies: either being absorbed or being eliminated. The applied sciences have always been allied to capitalism; they drive technology and flood the market with products. The pure sciences, or those with non-industrial applications, for the most part have been revised in terms of their driving principles; they have been adapted to and absorbed by the needs of our times. Present-day utilitarianism revolts at the idea of knowledge for its own sake. Even Buddhism, with its initially anti-materialist ideas, has been turned into a marketable product in the book shops or in the form of courses on transcendental meditation. Culture has also been turned into a "culture industry", to use Horkheimer and Adorno (1947)'s expression.

Genius science is substituted with science of the masses and for a democratic science that advances with the rhythm of mediocrity.[1]

---

[1] This book is about science, not about politics. Some colleagues who read a draft of this book warned me that I should not enter in political discussions, and even less so if I am going to adopt a position against the majority: One may flatter democracy in any context, but to disdain it is politically/academically incorrect and might prejudice my discourse on the sciences, with the possibility that my book might be rejected. A terrible reality: democracy gives freedom to individuals provided that they do not use it. Anyway, whatever my ideas about politics are, I shall not discuss them here. The word "democracy" appears several times in this book but that is not done as the fancy takes me. It is done because there is a strong connection between the way in which science is done and the dominant socio-economical scenario. For instance, as noted, the manner in which the quality of a paper is evaluated, counting the number of citations, indicates a

The stomach is put before the brain. Everything is bureaucratized, everything requires paperwork and to conform to mediocrity in order to effect a project.

"Democratic resentment denies that there can be anything that can't be seen by everybody; in the democratic academy truth is subject to public verification; truth is what any fool can see. This is what is meant by the so-called scientific method: so-called science is the attempt to democratize knowledge - the attempt to substitute method for insight, mediocrity for genius, by getting a standard operating procedure. The great equalizers dispensed by the scientific method are the tools, those analytical tools. The miracle of genius is replaced by the standardization mechanism". (Norman O. Brown, 1966)

Finding experts in a subfield of science who know nothing, or nearly nothing, about general culture (history, geography, literature, philosophy, sciences distinct from their specialization) is very common. It is sufficient to examine how they behave, how they dress, what their interests are, to realize that they are mostly quite ordinary people. There is not much difference between the average senior scientist and a seller of melons in the market. Both run a business to make money and to allow them to live with their spouses and their children in the welfare state. Henpecked husbands who carry out research in their domestic science during office hours, trying to reach agreement with their adjacent colleagues, thinking about school for their children or other domestic affairs rather than concentrating on daring scientific problems. Senior male scientists wearing bermuda shorts and T-shirts in the workplace, some of them with rings in the ear or in the nose, or a tattoo. Egalitarianism has, with few excep-

---

democratic criterion: What most people cite must be the best thing. There are also perfect examples of democratic science when one hears, for instance, that whether or not the neutrino has got a non-null mass was decided by voting among the specialists, what most of them voted being taken as correct These are not two separate things, and a thinker who is not a mere specialized bureaucrat should worry about the connections of science, politics, history and many other areas. We cannot talk about the sociology of science without discussing other realities away from science. This is the main reason for mentioning "democracy" herein, but, I repeat, this is not a book about politics, and I do not try here to argue in favour or against any form of politics.

tions, converted all the intellectual classes in plebeians with the same habits. Scientific knowledge has become a milk cow on which to grow fat,[2] an industry providing jobs for state employees. The function of science nowadays is far from the function science had in the times in which it was considered to be a source of wisdom.

Research institutes are becoming more interested in social rights than in the search of new, illuminating truths. For instance, there is a concern about having an equal ratio of women/men in all fields of research, or some posts for disabled people, etc. Certainly, there are good scientists who are women and they should have the same rights as men to gaining access to research centres, but absurdity arises when some institutes reject the good CVs of some men in favour of more mediocre women, just because they want a higher representation of women, just to maintain the external image of parity. The same thing can be said about disabled people or other under-represented sectors of the population. There are in Spain now laws which oblige to have a minimum representation of 40% of women in all scientific committees. Absurd! I have also observed how the publicity about women in research is organized. For marketing purposes in campaigns about women in science, a mediocre woman is usually selected to show in the mass media just because she represents what common people would like to believe: a young, attractive woman, a good spouse and mother, doing science from nine to five, after leaving the children in the kindergarten, the image that many women want to see reflected. There are however some older women, with many wrinkles and grey hair, who are excellent scientists, consumed by their long career, who have spent more time in laboratories than in beauty salons and in family life, but they do not offer the image that people want to see, and they are not very visible in mass media.

The fact that science has become a big enterprise, consuming huge amounts of state funding has made it more vulnerable to becoming politicized and subject to the social values of the masses rather than the values of a thinking elite. Not only are scientists at the service of mediocre programs of research devised by mediocre scientists who dedicate more time to bureaucracy and getting funds than to thinking about science; we have even reached the incredible situation that ordinary people without much idea about science are

---

[2] "Science. Heavenly goddess for some people, and an industrious cow which produces butter for other people". (Goethe)

being asked to propose topics for the future direction of science. The argument is that people pay their taxes and scientists use part of these taxes to do their science, so people have the right to choose which projects the public money is invested in. See, for instance, the web page http://www.reto2030.eu/, where people may vote for which scientific projects should be financed before the year 2030. Not only mediocre scientists choose, now non-scientists also choose.

Not only do researchers spend most of their time in administrative tasks, but there is also a huge number of posts in research institutes or university departments which are dedicated fully to administration, bureaucracy, filling in forms. Spain, my country, is one of the most representative examples in this aspect. In the institute in which I work in Tenerife, there is a ratio of one administrative post for every 2-3 researchers (including staff, postdocs and Ph.D. students). One might think that with that a heavy load of staff dedicated to administrative tasks, researchers would be free of those tasks and even have help to type their papers, fill in their forms, etc.; but no, that is not the case. Researchers must do most of the administrative work, and administrators are there just to stamp the completed forms and ... to take coffee in the cafeteria during working time. That is Spain, and the situation is not the same in all countries, although it is not a unique case. In Italy, for instance, as noted in the section *"Il trionfo della burocrazia"* by Sylos Labini & Zapperi (2010, ch. 4), of the 6,000 undefined contract positions at CNR (Conseglio Nazionale delle Ricerche) in 2006, only 60% were researchers, while the other 40% were administrative (mainly) and technical staff.

This is the situation in contemporary culture, not just in science. We must admit that the idealistic approaches have lost the war. In the end, pragmatic industrialization of culture, bureaucracy and mediocrity have triumphed.

6.2   The Decline of Universities

Universities in general do not have future prospects any better than those of the pure research centres. Several factors contribute to this decline: The massive influx of people into universities, the progressive decrease of standards in taught subjects and the preservation of culture, the increase of bureaucracy and pragmatism, plus all the problems associated to research.

Let us select some paragraphs of the manifesto about the decline and fall of the British university by Mark Tarver (2007), an exlecturer in a School of Computer Studies, who left his job because he became fed up with so much decadence. The context of this text is the British university, and he talks about Computer Sciences, but I think many of his arguments may be applied to the university in general and in most countries nowadays, at least in Europe and US, and in any field:

> "An immediate casualty were some hard-core traditional CS modules like complexity and compiler design. Why, argued students, elect to study some damned hard subject like compiler design, when you could study something cool like web design and get better marks? So these old hard core subjects began to drop off. Even worse, the School (following the logic of the market), having seen that these hard core subjects were not attracting a following, simply dropped them from the curriculum. So future students who were bright enough to study these areas would never get the chance to do so. After a few years of this system, the results percolated through to my office. I could see the results in the lecture hall, but the procession of students who walked into my office and said 'Dr Tarver, I need to do a final year project but I can't do any programming' [...]
>
> Graduating computer-illiterate students who had to do a project in computer science was more of a headache. The solution was to give them some anodyne title that they could woffle or crib off other sources. It was best not to look too closely at these Frankensteinian efforts because otherwise you would see stitches where they lifted it off some text which you were never likely to find short of wiring them to the mains to get the truth. It was of course, a lie, but the cost of exposing that lie was likely to have ramifications beyond the individual case. Very few lecturers would want to stir such a hornets' nest or have the necessary adamantine quality to inflict shame upon a student whose principal failure was to be allowed to study for a degree for which he had little ability.
>
> After seven years of the new regime, I had the opportunity to compare the class of 1999 with the class of 1992. In 1992 I set a course in Artificial Intelligence requiring students to solve six exercises, including building a Prolog interpreter. In 1999, six exercises had shrunk to one; which was a 12 line Prolog program for which eight weeks were allotted for students to write it. A special class was laid on for students to learn this and many attended, in-

cluding students who had attended a course incorporating logic programming the previous term. It was a battle to get the students to do this, not least because two senior lecturers criticised the exercise as presenting too much of a challenge to the students. [...]

Now parallel with all this was an enormous paper trail of teaching audits called Teaching Quality Assessment. These audits were designed to fulfil the accountability of the lecturers by providing a visible proof that they were doing their job in the areas of teaching and (in another review) research. In view of the scenario described, you might well wonder how it is possible for such a calamitous decline in standards to go unremarked. The short answer is that, the external auditors, being lecturers, knew full well the pressures that we were facing because they were facing the same pressures. They rarely looked beyond the paperwork and the trick was to give them plenty of it. The important thing was that the paperwork had to be filled out properly and the ostensible measures had to be met. Students of the old Stalinist Russian system will know the techniques. Figures record yet another triumphant over-fulfilment of the five-year plan while the peasants drop dead of starvation in the fields.

Teaching was not the only criterion of assessment. Research was another and, from the point of view of getting promotion, more important. Teaching being increasingly dreadful, research was both an escape ladder away from the coal face and a means of securing a raise. The mandarins in charge of education decreed that research was to be assessed, and that meant counting things. Quite what things and how wasn't too clear, but the general answer was that the more you wrote, the better you were. So lecturers began scribbling with the frenetic intensity of battery hens on overtime, producing paper after paper, challenging increasingly harassed librarians to find the space for them. New journals and conferences blossomed and conference hopping became a means to self-promotion. Little matter if your effort was read only by you and your mates. It was there and it counted.

Today this ideology is totally dominant all over the world, including North America. You can routinely find lecturers with more than a hundred published papers and you marvel at these paradigms of human creativity. These are people, you think, who are fit to challenge Mozart who wrote a hundred pieces or more of music. And then you get puzzled that, in this modern world, there should be so many Mozarts—almost one for every department.

The more prosaic truth emerges when you scan the titles of these epics. First, the author rarely appears alone, sharing space with two or three others. Often the collaborators are Ph.D. students who are routinely doing most of the spade work on some low grant in the hope of climbing the greasy pole. Dividing the number of titles by the author's actual contribution probably reduces those hundred papers to twenty-five. Then looking at the titles themselves, you'll see that many of the titles bear a striking resemblance to each other. 'Adaptive Mesh Analysis' reads one and 'An Adaptive Algorithm for Mesh Analysis' reads another. Dividing the total remaining by the average number of repetitions halves the list again, Mozart disappears before your very eyes.

But the last criterion is often the hardest. Is the paper important? Is it something people will look back on and say 'That was a landmark'. Applying this last test requires historical hindsight—not an easy thing. But when it is applied, very often the list of one hundred papers disappears altogether. Placed under the heat of forensic investigation the list finally evaporates and what you are left with is the empty set.

And this, really, is not a great surprise, because landmark papers in any discipline are few and far between. Mozarts are rare and to be valued, but the counterfeit academic Mozarts are common and a contributory cause to global warming and deforestation. The whole enterprise of counting publications as a means to evaluating research excellence is pernicious and completely absurd. If a 12 year-old were to write 'I fink that Enid Blyton iz better than that Emily Bronte bint cos she has written loads more books' then one could reasonably excuse the spelling as reflective of the stupidity of the mind that produced the content. What we now have in academia is a situation where intelligent men and women prostitute themselves to an ideal which no intelligent person could believe. In short they are living a lie".

Another testimony to the decline of the university, this time in Spain and referring more to humanities, comes from Juan Arana (2003). Although this author warns that his conclusions are only from his personal experience as Professor of Philosophy in a Spanish university and possibly not applicable in the same way to the science faculties, I think that some of his statements are useful observations about the university in general. Here I translate into English some of the paragraphs of his paper in Spanish:

"Some years ago, a group of students asked me for a grant to publish a journal: They were told that the funds of the culture department were already promised to pay for a concert of rock music and some barrels of beer. [...] In other ways, we are attending a bargain sale, in which titles, exams, marks, academic credits are becoming cheaper and cheaper. However, the public is becoming progressively meaner and meaner: There is no way that people will attend conferences without the ratification of classes; go to congresses, if there is no grant for attendance; to seminars, concerts or series of club-cinema, if they are not paid with credits. [...] the 'level' has to be revised downwards every so often. The absenteeism of students is becoming more frequent; their incapacity or lack of motivation to read books (even the subject's set book) is more frequent too. Their spelling errors are no longer a matter of scandal. [...]

The emergence and rapid development of evaluation agencies of different kinds and jurisdictions ensures that university life develops in a narrower and narrower climate of negotiations which leave aside the intrinsic quality of the work and give priority to purely formal requisites, such as the development of reports, the completion of forms, creation of information dossiers, visits of evaluators, discussion sessions, etc. etc. etc. My own experience, contrasted with that of some of my colleagues, is that all this is useful only to hugely increase the administrative tasks and favour the experts in bureaucratic trickeries [...]"

This is certainly not the case in all countries, but to some degree many of these observations are generally applicable. Summing up, the level and quality of university education is decreasing while bureaucracy and administrative tasks are increasing. The vulgarization of the academies is already a fact, and the relative amount of people who do not care about culture in universities is constantly increasing. Other negative circumstances are also associated with the university. For instance, the hierarchical structure in Italian universities is comparable to feudalism (Sylos Labini & Zapperi, 2010), with too much power concentrated in the older generations, leaving young researchers in precarious conditions. The power structures, the endemic growth of departments, the dependence on politicians, the corruption, etc. are also problems in academic structures.

The consequences of this directly affect the quality of scientific research. The new generations are, on average, worse prepared than older ones to solve problems with intelligence. For instance, mathematical ability in theoretical physics is being progressively substituted by the ability to do computer simulations. One might think that this is a consequence of the advance of science, which has reached the point where solutions can only be explored with computer calculations. This is partially true, but it is also true that many theoreticians have not learnt how to think about a problem without a computer. It is like contemporary artists who are not familiar with the techniques of the classical fine arts. Indeed, the problem is not whether calculations are carried out with computer or analytically, the problem is that scientists are losing their capacity to think, and they are more like machines which press computer or instrument buttons. Intelligence and a higher capacity for thought are becoming extinct within science and tend to be substituted by technology, and the bureaucratic tasks of filling in forms to get money and use that technology. Am I exaggerating? Yes, I am. The situation is not at present so extreme, but the trend is towards that.

6.3   (Lack of) ethics in science

Contemporary science is neither more nor less moral than science at other times. Cheating, plagiarizing, corruption, etc. is part of humanity in all epochs and places, although to different degrees. And vice versa: honest scientists have always existed as they exist today, as the majority. Nonetheless, the fact that science is becoming more like an industry, with economic interests, greater competition, the less idealistic approach to science, the power of the mass media, the way science is funded, etc. all contribute to malpractice. Among the typical misconduct, we can find: falsification or manipulation of data; publishing a paper knowing it is wrong; hiding information from a paper when it is asked for because of the suspicion that something might be wrong; referees/editors who reject correct papers or delay their publication with false arguments because of a conflict of interests; the same for reports on funds distribution, use of instruments; plagiarism or not citing original sources of some ideas, or obstructing knowledge; abuses of authority by monopolies (e.g., *arXiv.org*, a preprint server in physics and mathematics; see criticism in Castro Perelman, 2008a); etc.

After enumerating a large number of cases of lack of ethics in science, Bauer (2008) says:

> "I hope you agree that all this is unpleasant, sleazy, and shouldn't happen. But does it have anything to do with the actual science? Does it really matter, who gets the credit, so long as science keeps progressing?
>
> I think it does matter—because science progresses with sound, reliable results only to the degree that scientists are honest.
>
> Most people think science gives trustworthy results because of "the scientific method": testing ideas by experiment and so either proving or disproving them. Isn't that what you all do? Experiment, and find out what's true and what isn't?
>
> But what if an experiment doesn't give the result you expected? What if it gives a result that you just know is wrong in some way? Don't you keep trying until you get the "right" result? Especially if you know that your boss is very sure that's what you should get? Isn't there the temptation to fudge a bit? Since you know what the right answer ought to be, why not just round the numbers off a bit?"

And Bauer (2008) concludes his paper:

> "Faculty evaluating others for tenure or promotion; administrators deciding how to calculate overhead charges, and how to distribute money collected as overhead; program managers in funding agencies; scientists reviewing research proposals and manuscripts intended for publication—in all those situations and many others, we as individuals have to decide how much weight to give to sheer merit as opposed to other considerations like scratching one another's back. If most of us choose the ethical thing, then science will continue to prosper. If too many of us cut corners, then science can come to a dead stop".

Bauer points out one of the possible factors which contributes to the decline of present-day: lack of ethics. However, we may consider, as noted, that this is not exclusively a problem of our time, and also that there are some mechanisms for punishing people's misdemeanors. Some authors have asked for a panel of ethics in science; for example:

> "All we can do is grasp the cliché: hope for the best and be prepared for the worst. This is the way life is, some will say, and nothing can be done. If the genes are so unjust, how do we expect to have a just world to begin with? This is the question that has been haunting biologists for a long time. Whatever our most impending fears of the future might be, what is certain is that a worldwide scientific panel of ethics is desperately needed, where scientists may address all their grievances and they themselves can be brought to justice and be held accountable for their misdeeds. As far as oxymorons go, 'ecological capitalism' is one of them. Let's hope that 'scientific ethics' is not another one". (Castro Perelman, 2008b)

And indeed there are in certain countries committees to evaluate the faults of the active scientists in their profession. Do they work? That is the main question to ask.

Let's pay attention, for instance, to the cases of falsification/ manipulation of data, which have appeared in published results which are eventually discovered to be failures. The question is not simple, because there are failures that are unconscious lapses, or botched work done without care, and failures which are conscious cheating, and it is difficult to distinguish them because the authors do not usually admit to cheating. Salpeter (2005) has pointed out that both in astronomy and medicine, fallacies do occasionally occur. Sometimes these fallacies are due to "subconscious cheating" or the "file drawer effect", where favourable results are published but unfavourable or inconclusive results are merely filed away. But apart from these "well intentioned" errors, there are also cases which can be classified as "conscious and intentional cheating", and there are in the middle a large number of cases of conscious cheating done with good intention, because the researchers are very convinced about what should be the true result of their experiments and pay more attention to their prejudices than to the numbers given by the measurement devices.

The historical example of the measurement of the electron charge is very well described by Feynman (1974):

"We have learned a lot from experience about how to handle some of the ways we fool ourselves. One example: Millikan measured the charge on an electron by an experiment with falling oil drops, and got an answer which we now know not to be quite right. It's a little bit off, because he had the incorrect value for the viscosity of air. It's interesting to look at the history of measurements of the charge of the electron, after Millikan. If you plot them as a function of time, you find that one is a little bigger than Millikan's, and the next one's a little bit bigger than that, and the next one's a little bit bigger than that, until finally they settle down to a number which is higher.

Why didn't they discover that the new number was higher right away? It's a thing that scientists are ashamed of—this history—because it's apparent that people did things like this: When they got a number that was too high above Millikan's, they thought something must be wrong—and they would look for and find a reason why something might be wrong. When they got a number closer to Millikan's value they didn't look so hard. And so they eliminated the numbers that were too far off, and did other things like that".

Feynman (1974) also says in another part of his paper:

"It is interesting, therefore, to bring it out now and speak of it explicitly. It's a kind of scientific integrity, a principle of scientific thought that corresponds to a kind of utter honesty—a kind of leaning over backwards. For example, if you're doing an experiment, you should report everything that you think might make it invalid—not only what you think is right about it: other causes that could possibly explain your results; and things you thought of that you've eliminated by some other experiment, and how they worked—to make sure the other fellow can tell they have been eliminated".

I guess this is also the history of cosmological parameters, or in many other fields in science. But I also think that present-day science is not much worse than past science in this aspect. Of course, the errors are human, and the beliefs in a "preferred result", or the fear of being considered as an outsider when giving numbers different from the consensus agreement are part of the guilt. Anyway, it is lack of objectivity in the scientific method and, even worse, a lack of ethics. Nothing can justify the intentional falsification/manipulation of data; and also the file drawer approach should be considered as misconduct of science. The individuals are not the only guilty parties, but the whole system which rejects people whose results do not match the general consensus. The system is pressing scientists to get results which agree with those of the establishment, either directly, or by twisting/cheating data to fit standards, or else filing results away.

Solutions to this problem may be considered. We can create audits to prosecute all suspected cases of conscious cheating and apply penalties to them; but this, if carried out strictly, would transform science in witch hunts where everybody suspects each other, and it would be a waste of time and money. Or we could turn a blind eye to them; allowing an increasing corruption in science. Or we could prosecute and punish only those cases which cause direct damage to somebody else; or we could cancel all economical benefits obtained with the failures (grants, tenures, funds) or even ask for money to be refunded. Perhaps, the cheapest and most efficient punishment, and the most usual, is the discrediting through gossip of a scientist who cheated; but this is not fair since much gossip is a hoax.

There are real cases of judgements and penalties. For example, the case of the Tsukuba University (Japan) against its researcher, Teruji Cho, and his collaborators in plasma physics. The Investigation Report on the Suspected Scientific Misconduct, by "The Investigation Committee in the Scientific Ethics and Research Conduct Committee at the University of Tsukuba", November 26, 2007, concluded that Figs. 1a and 3 of the paper by Cho *et al.,* (2006) reveal a falsification of data:

"Fig. 1a: the S/N ratio of the raw data turned out to be quite low, while the data of points in Fig. 1a show a smooth variation. The declared analysis procedures to obtain such smooth data points with small errors from raw data of low S/N ratio lack objectiveness and scientific basis". "Fig. 3: the data from different shots are mixed in figs. e-h. The procedure called 'offset' employed in the data analyses for figs. a-j lacks scientific soundness. Furthermore, no reliable response was provided as to the identification of the raw data for figure i".

Consequently, on 4 March, 2008, Cho et al. were advised to withdraw and retract their paper, Cho et al. (2006). Teruji Cho and some of his colleagues decided instead to send an "Erratum" to the journal specifying the errors in their Figures 1a and 3. The University Committee considered that this was not enough and, on 27 August, 2008, adopted a harder measure: Cho was dismissed from his permanent position at Tsukuba University and three assistant professors, collaborators on that paper, were suspended for 1-4 months. After that, there were several responses, attacking or defending the University Committee's decision (Berk, *et al.*, 2008; Mizubayashi, 2009; Cho, 2009; Berk & Fisch, 2009; Akahira & Mizubayashi, 2009) and a civil suit will take place in which Teruji Cho makes a claim again the university (still without resolution at the time of writing).

As we can see, Japanese people take things very seriously and do not balk at applying drastic penalties, but could we imagine a world in which any suspected case like this might be converted in a long debate lasting years, with scientific and civil suits, etc.? I think we would waste more time in these struggles than in doing science. Lawyers would benefit from this process and a lot of new rules and laws on scientific practice would be created. It is typical of a decadent scenario. Decadence and the law have a certain correlation in society. It has well known for a long time by learned people in China, thanks of the experience obtained from the history of that country, that when empires are falling down, they tend to have many laws. The Colombian thinker Gómez Dávila (post. 2007) said: "Moribund societies accumulate laws like moribund people accumulate remedies".

If we are going to talk about lawyers and decadence, thinking about the US is unavoidable. A country which is tailor-made for justice professionals, in which a large portion of the economy is generated through civil suits, and most professionals need a lawyer and legal insurance against claims. The paradise of the pettifogging lawyers! A heavy load which may sink a country… But this is not the topic here. Rather, I will comment on some adjudicated cases of fraud in scientific research.

The National Institutes of Health (NIH) in U.S. have a fraud unit which is responsible for judging possible frauds in topics related to medical research. Let us examine two cases considered by this unit:

1. The case of Mark Spector (Racker, 1989) is a good example of how easy is to apply penalties to a weak person. Mark Spector was a Ph.D. student. He isolated a new protein kinase activator by transforming growing factor in tumor cells. However, radiactive iodine (an old-known protein activator) was later discovered in Mark's

sample so the new activator discovery was indeed a failure. Mark replied that somebody else had put iodine in his samples, but apparently his defence was not convincing enough. In the end (1981), misconduct was claimed, his papers were withdrawn, his Ph.D. withheld, and he could not continue his career as researcher (he moved to South America).

2. The case of David Baltimore (Hilts, 1992) is a good example of how difficult it is to apply penalties to a high-status scientist. David Baltimore received a Nobel Prize in Medicine in 1975. In 1986, he and his colleagues published some results of work on antibodies/immunology (Weaver, *et al.*, 1986). A young postdoc working in his team, Margot O'Toole, failed to reproduce their results and realized that some of the work cited in that paper had not been carried out, so she suggested to her bosses that they withdraw their paper, and denounced the case. The case was against one of the coauthors, Theresa Imanishi-Kari, but was linked to Baltimore's name because of his scientific collaboration with and, later, his strong defence of Imanishi-Kari against accusations of fraud. Baltimore attacked O'Toole as a discontented postdoc and wrote letters to hundreds of colleagues, claiming that the NIH misconduct investigations were a threat to science itself, casting the conflict as one of "outsiders invading the sanctuary of science". (What a sentence to remember!). Baltimore et al. spent tens of thousands of dollars (from his institute) on lawyers and in the end (1996), the government failed to prove any of the nineteen charges. The *New York Times* in its editorial on 25 June, 1996, called the case "The fraud case that evaporated".

We cannot know whether the NIH fraud unit's conclusions were correct or not, and we cannot judge the judgers. However, it is well known that, in science and in other fields, justice is usually corruptible in favour of the powerful people or biased towards money-owners. Is the solution (the corruption in the board judging a case) better than the problem (the corruption in science itself)? Possibly, yes; at least most scientists will be afraid of penalties and will reduce the trend to manipulate/falsify information.

6.4   The Perelman case

Grigori Yakovlevich Perelman (1966-) is a Russian mathematician, who has made landmark contributions to Riemannian geometry and

geometric topology. In particular, he proved Thurston's geometriza-
tion conjecture. This solves the Poincaré conjecture, posed in 1904,
which was viewed as one of the most important and difficult open
problems in topology until it was solved. In August 2006, Perelman
was awarded the Fields Medal (equivalent in prestige to the Nobel
Prize for Mathematics) for "his contributions to geometry and his
revolutionary insights into the analytical and geometric structure of
the Ricci flow". Perelman declined to accept the award. On 22
December, 2006, the journal *Science* recognized Perelman's proof of
the Poincaré conjecture as the scientific "Breakthrough of the Year",
the first such recognition in the area of mathematics (MacKenzie,
2006). He has since ceased working in mathematics. On 18 March,
2010, it was announced that he had met the criteria to receive the
first Clay Millennium Prize Problems award of US $1,000,000 for his
resolution of the Poincaré conjecture but on 1 July, 2010 he turned
down this million-dollar prize, saying that he believes his contribution in proving the Poincaré conjecture was no greater than that of
US mathematician Richard Hamilton, who first suggested a program
for the solution (Wikipedia).

Perelman is not a rich man. Rather, he is an ascetic, living hum-
bly with few resources. The question then is why he has rejected
those two very important prizes. And the answer was given by
Perelman himself (Ritter, 2010): "the main reason is my disagreement
with the organized mathematical community. I don't like their
decisions, I consider them unjust". The mass media throughout the
world has pointed out the eccentricity of his behaviour. A portrait of
a "crazy genius" became the most popular image for the explanation
of his behaviour. Perhaps this is the most convenient explanation for
the mathematical community: a person who rejects a million dollar?
A person who rejects the most important prize in mathematics? He
is a crazy man, a scientist who lost the plot. There is however another possible interpretation: Perelman is a hero, a martyr, and with his
sacrifice he wants to denounce the misconduct within the mathematical community. Possibly, a combination of two facts is the real
situation: a crazy hero.

Indeed, this is a revolutionary act, one of the most amazing revolts in science of the last few decades. We can leave aside all the
press releases of the media, conveniently oriented towards a politically correct interpretation. We can observe the facts and draw our own
conclusions. My conclusion would be something like this: the most
important mathematician of our time is claiming that the mathemati-

cal community works in a very decadent way, and he refuses to be part of the farce. He rejects the money and the prizes because, if he accepted them, he would be considered to be within the system, but he wants to spit at the system from outside. He wants to raise the flag of the unfairly rejected outsiders by joining them, although he was a professional mathematician, not an amateur. He may be an extravagant character, an atypical individual, in rejecting huge amounts of money, but he does not look a fool. Possibly he has calculated very well his movements. His rejection of the prizes is likely to be his revenge, possibly because of many bad experiences with referees, journals and the mafias within the establishment.

It is more convenient for the establishment to think about the melancholic Perelman as an ascetic man, refusing to engage with the rest of the world and playing violin while living with his mom. Because, if we were to think about the real meaning of his rejections, according to his own words, the establishment would have to worry about its role and dignity. If there are outsiders who are better mathematicians than the establishment mathematicians, one might starting pondering the dangerous idea: what are the scientific institutions for? Would science not be more convenient without its institutionalization? Given that people like Perelman, the greatest mathematician, does not need the support of the institutions, what about closing all mathematical research institutes or finishing all funding for mathematics? Very dangerous ideas for the professional administrators of science.

Mathematics is a special kind of science and the conclusions about it should not be generalized to all sciences. In this present book, I am more interested in natural sciences. We may wonder though whether some of the conclusions of this case might be extrapolated to the rest of the sciences. My interpretation is that a case like Perelman's is less likely to be repeated in the empirical sciences because the demonstrations of a hypothesis are less objective and more open to interpretations and prejudices than the demonstration of a theorem in mathematics. In mathematics, if a genius discovers an important truth within the world of mathematical ideas, this is undeniable and cannot be rejected. In physics, biology, and other natural sciences, if a genius discovers an important truth about nature, this might be unconsidered or rejected by the establishment as an unproved speculation (and to make inconvenient changes in the worldview of the establishment no evidence is extraordinary enough). There might exist even now some cases equivalent to Perelman's in natural sciences with regard to professional scientists who have solved important scientific problems but not been recognised. Anyway, surprises are not discarded: perhaps a politically correct scientist nominated for the Nobel Prize in Physics might reject the prize on the grounds that the physics community misbehaves. It would be a noble and brave attitude, although rare in a person (normally an old person) who has spent most of their life within the establishment. In other fields, there have been some rejections of the prize, such as the case of Jean Paul Sartre for literature; but writers/humanists are more individual creators, and it would be strange for a renowned scientist to behave in that way. There are however cases of Nobel Prize-winners who use their status to challenge the unfair decisions of the establishment; for instance, Brian D. Josephson (1940-). Regrettably, this last case is related to studies of paranormal events, claiming that parapsychological phenomena may be real, and that Eastern mysticism might have relevance to scientific understanding, which is enough of a reason, in my opinion, not to take him very seriously. It does not alter the fact, however, that his observations on the sociology of science and his works in condensed matter physics might be sound.

6.5   Chronicle of a Death Foretold

Chronicle of a Death Foretold (1981) is the English translation of the Spanish title of a novel by the Colombian Gabriel García Márquez, winner of the Nobel Prize for Literature. It might be also be applicable to the state of science or other branches of culture as well. At present, we do not know how this death will occur, so we may speculate on some ideas related to it. The spirit of science is already languishing, but the structure of scientific research is still very alive. Tentatively, we may foretell a possible way in which science will reach its institutional death.

Since the goal of science as an institution is mostly socioeconomic—keeping a structure which creates employment for myriad members of the guild, and allowing some people to acquire some power—the evolution of scientific knowledge will not directly affect its existence. The problem for scientific institutions will come when its influence over society is reduced and when the resources that science consumes begin to diminish.

One possible reason for stopping the expansion of scientific investment and causing its collapse is that science will reach the maximum expense that a society can afford. Scientific institutions follow the structure of capitalism, so they must continuously grow. Experimental science becomes more and more expensive with time, and science has opted for this way of no return, going always for an increase in funds. When a limit is reached at which the investment in science can grow no more, a crisis will become unavoidable. Nowadays, the richest countries invest around 3% of GDP in research and development, from which 20% is for pure sciences, a ratio much higher than in the past, both in absolute and relative terms, and that has grown continuously in the last few decades, with some small fluctuations. Possibly this is already close to the asymptotic limit in terms of the relative ratio of money that a society can afford, so a crisis may be not very far away. The crisis will also depend on circumstances in society: if the GDP of developed countries grows through inflation, even at a constant ratio it will result in an increase of investment in science; the exponential fast growth of the last few decades cannot be sustained but at least a slow growth may delay the death of science.

As I said in Chapter 1, society may soon realize that science cannot satisfy its expectations, and that the returns from hyper-million investments become smaller and smaller, so it is possible that the richest countries will reduce the titanic economic efforts necessary to produce tiny advances in science, at least within the pure (non-applied) sciences.

In developing countries, the ratio of GDP dedicated to scientific research is quickly increasing, and this may sustain science in the global world for some time. One of the arguments for promoting an investment in science in developing countries is that there is a correlation between wealth and investment in science, so the governments of these countries are persuaded that a higher investment in research will contribute to economic development. However, they will soon realize that a correlation between A: investment in science and B: wealth does not necessarily mean that A is the cause of B; rather, B is the cause of A: the richest countries invest much more money in science because they can, it is a luxury in countries with more than enough resources. When developing countries realize that investment in science does not produce the economic returns they expected, they will reduce these investments too.

In this pessimistic scenario, the collapse of foundational science is very likely, at least of experimental science; and since advances in theoretical science have become totally dependent on the advances in experimental science, this will also produce a collapse in the supply of cheap theoreticians over a longer period of time. The effect will not be immediate. Possibly, many centres will continue for some decades with a decreasing budget, but eventually they will recognize that no advances can be made with small budgets; even less than the few ones obtained with the huge budgets they have got nowadays. Therefore, research centres will begin to close, one after another.

I guess applied science will last much longer, but without the support of research in pure science, this will be transformed into mere engineering. This is equivalent to science death. Indeed we can already observe some symptoms of this decline of science: Many scientists working on research within pure science dedicate most of their research time to work, apart from administrative tasks, as engineers: development of instruments, computer pipelines, etc. The reason is simple: there is more money for technical applications, and modern-day scientists move to earn money and power.

Medicine is perhaps a special case of science whose future lies away from that of other sciences such as physics, astronomy, biology, etc. Humanity's obsession with living a long life, without illnesses, the illusion of eternity, makes the pursuit of better health an endless race. Fear of death remains, as with religion, the main focus of this science, and while this focus remains, the institutional body will also remain. Indeed, all societies have some type of medicine, though some of them work within a closed body of traditional knowledge and have no interest in the search of new techniques. Nonetheless, the increase of the costs of medical research may put some constraints in the development of this area as well.

Another factor to take into account is the possible lack of interest in younger generations in dedicating their time to research. On the one hand, we see that scientific work is becoming more and more like a business or an administrative task; it is becoming a boring office job rather than an exciting adventure. On the other hand, there are other jobs which require high qualifications and they are as boring as present-day science but with higher salaries, so many young people with higher education will prefer to invest their time in these other jobs and the vocation for science will be considerably reduced. This is a consequence of the capitalist scenario in which new generations are educated. We live in a competitive society in which the winner gets the reward of a higher salary or status. Most people go to university thinking that the degrees they obtain will be useful to earn higher salaries, so it is a logical consequence that people try to invest their educational time to maximize the economic benefits. Also, the reputation of a scientist or any intellectual profession was one time highly considered, but nowadays tends to be considered at the same level as engineers or as just a technical profession. People forget that a scientist is or should be an intellectual, not merely a technician, and our society is moving quickly towards a devaluation of the intellectual and "culture for the sake of culture", which are being replaced by utilitarianism, light culture for the masses, and culture as business or a fun fair for tourists. Hence, people will see the career of the scientist as a big effort for small rewards, and they will prefer other options. These problems are similar to those of the Catholic church, suffering from a lack of vocation in Europe. Possibly, like the church, science can recruit people from developing countries, but the success of this recruitment will depend on how much money science is able to offer as salary, because the major goal of most highly educated individuals from poor countries is to move out themselves and their families out poverty. Of course, there will be a few young people everywhere with the true vocation of scientists, who will only want to do research in science, but the amount of manpower necessary to keep the present-day machinery of science going will be significantly reduced and the structures will be very much affected.

Without money and without people, the prestige of science will be much eroded. It will be converted into a residual cultural activity of the past, something similar to what happens nowadays with philosophy, which at one time was of great influence in society—many kings were indeed interested in philosophy and asked advice of philosophers, for instance—, and now lives in frustrated isolation, put in an academic corner where it cannot affect the way society works.

A crisis in the business of science, a crisis of no return, will happen, and a new dark age of scientific knowledge will arise. This will not happen very fast but will be a slow process, possibly lasting several generations, and this decline will not only affect science but the sinking of science will run parallel to the sinking of many other aspects of our civilization. Indeed, they will most likely feed off each other. Science is a major characteristic of our western culture, and our way of thinking. Therefore, the end of science will mean the end of modern European culture, the twilight of an era initiated in Europe around the fifteenth century and which is extended nowadays throughout the world: the scientific age. Will be there a new renaissance after the dark ages? A new dawn some centuries after the twilight? Will there be a replacement? Who knows?

The Decline of the West is the title of the classic masterwork by Oswald Spengler, about which I will speak in the next chapter, a book which discusses a state of things which is now more evident than when the book was published nine decades ago. My message here is similar in some ways, but I just pay attention, in this book, to the world of science. Nonetheless, human culture is interrelated and one may expect that the symptoms of the sickness of a civilization will be manifest in different parts of the social body in an almost simultaneous process, with only some short delays with respect to each other. Decadence in art seems to be the first visible symptom of the decline of our civilization, with clear signs of an "end of art" era since the beginning of the twentieth century. Science may be one of the last cultural manifestations to topple. And between arts and sciences we have all the intermediate expressions of culture within the humanities, with a partly artistic/spiritual/subjective approach and a partly scientific/objective approach. Why should science be the last one to sink down? I guess that it is because it is less affected by the lack of meaning in the development of our civilization. Meaning is crucial for the arts. However, science may continue to be developed even without any relevance, in a mechanical way, like a robot in a factory, for some further decades. Anyway, the lack of meaning, due to excess useless knowledge, the bureaucratic power structures unaware of the search for truth and the diminishing returns with increasing costs, will eventually erode the organizations that produce science, and it will follow the same route as the arts. Art is produced by individuals while science may be produced in large groups. Individuals more quickly recognise the lack of meaning in an activity, while big groups behave more stupidly and need more time to realize the nonsense of their activities; this is also a factor in the swifter death of the arts. The science produced by isolated individuals, in a somewhat artistic way, will also die early, leaving space for macro-projects in science whose purpose cannot be clearly thought through by any individual scientist, and the motivation becomes just an economic affair, a way to distribute salaries among the employees of science. Art became dead economy. Science is now becoming a dead economic system. Today, science is as crushed as contemporary art. In the words by Feyerabend (1975): "Science failed to be a variable human tool to explore and change the world and rendered itself into a solid block of knowledge, impermeable to human dreams, wishes and hopes". When dreams die and only economic benefits are important, poetry is dead, art is dead, literature is dead, humanities are dead ... and eventually science will also die, buried by the weight of economic interests smothering human dreams.

7   PHILOSOPHIZING ABOUT SCIENCE

Natural science is a part of philosophy, but it is a discipline limited to the purpose of understanding nature and does not care about human affairs. Human affairs, as I have said in previous chapters, are very important for the development of science. Therefore, an approach to philosophy or the humanities in a broad sense (including psychology, sociology, history, politics, and others) seems quite reasonable in order to understand the problems of science nowadays. Indeed, it is probable that some will identify the present book as a philosophical criticism, a charge made by many philosophers against science. Certainly, I think that the way of thinking in this book is philosophical. However, it must not be confused with the type of presentations made by self-claimed professional philosophers. Philosophy of science done by scientists or philosopher-scientists is not the same thing as philosophy of science done by pure philosophers without direct contact with the profession of scientific researcher.

In my opinion, when talking about science or nature, listening to active scientists who produce their own philosophical reflections is the best option; and when talking about Philosophy with a capital P I prefer to listen to the great philosophers, that is, to the important classical philosophers rather than the mediocre specialized academicians of our own epoch. We may wonder whether the present-day philosophers of science may help science to be better, and my answer is negative; paying attention to them is a waste of time. Only a few excellent classical philosophers deserve to be read.

Saying that reading present-day philosophers of science is a waste of time has been an expensive sentence for this book. Because of this and other similar opinions, an anonymous referee (presumably a professional philosopher of science) who sent an evaluation of this book to an editor said: "We think this is not a satisfactory investigation into the topic the book pretends to cover. Just take a look at the summary of chapter 7, regarding philosophy of science: 'At present, professional philosophy... and reading contemporary philosophers is a waste of time'. What about [...]?"". Another anonymous referee (presumably, another philosopher of science), after being enthusiastic with the chapters in which I expressed hard criticism of science, they said: "However, what disturbs me considerably, is that the author appears to be a proponent of the traditional continental philosophy, and seems to consider the analytical (mainly Anglo-Saxon) philosophy to be trivial". As mentioned in Chapter 3 of this book, the role of peer reviewers is becoming more and more like a mafia, controlling which texts with ideas that differ from those of the dominant groups are rejected. This is more evident in philosophy, which is more subjective than scientific research.

The world of culture is divided into several mafias. If you want to say that present-day scientific research is a waste of time and money, you are not very welcome in scientific institutions, who fight hard to get state funds and are not willing to put those privileges in danger; however, you are very welcome in the departments of philosophy of science, where they earn their money with these kinds of stories. The opposite thing is also true: If you want to say in a research institute for science that philosophy of science is a waste of time, your message will be well received, with wicked smiles, but the people in the departments of philosophy of science will not like it and they will call you ignorant. What about the free thinkers who do not wish to represent any group? This is the case that I pretend to defend here: the right to be critical, independently of any mafia. Precisely because of that, I need to express clearly and without ambiguities that my role here is not to give more power to one guild, to the detriment of the power of another guild.

In this chapter, apart from a section dedicated to commenting on the mediocrity of so-called philosophers of science, I will provide some biographical notes and describe the work of some significant authors related to the subject of this book: Nietzsche, Unamuno and Oswald Spengler. There are many other philosophers with great ideas, but the present book is not an encyclopaedia, listing interesting thoughts throughout history. I suggest that readers consult other classical authors to find them. At present, I just want to introduce some ideas which may guide the discourse of this book, and these three philosophers have introduced some concepts which are worth knowing about. It may be noted that these names are not usually found in discussions about science, and only Nietzsche is well-known to the general public without a university education in philosophy. Moreover, these three authors belong to the tradition of continental philosophy (Unamuno is Spanish and the other two are German) rather than the authors born in or living in Anglo-Saxon countries who are usually cited. Also, the period in which these authors wrote is the end of the nineteenth century and beginning of

the twentieth century, rather than the usual referencing of late twentieth-century thinkers. Certainly, there is a different approach to
philosophy here, with respect to other authors talking about science.
As said by Einstein, "insanity is doing the same thing over and over
again and expecting different results", and I am not writing this book
to repeat again and again the same messages which fill the mouths
and pens of most academicians talking about it.

7.1   Do philosophers of science help science to be better?

"...we must confess that much of contemporary philosophy of
science and especially those ideas which have now replaced the
older epistemologies are castles in the air, unreal dreams which
have but the name in common with the activity they try to represent, that they have been erected in a spirit of conformism rather
than with the intention of influencing the development of science, and that they have lost any chance of making a contribution
to our knowledge of the world". (Feyerabend, 1970)

The philosophy of science as taught nowadays in the faculties of
philosophy is, indeed, a philosophy of anti-science. It is usually
taught that scientists are inept, unable to think about their own
science whereas the professional philosophers are those who are able
to give meaning to science, as well as establishing the limits of the
validity of science. There is a wide range of positions that could be
taken.

At one extreme, we have the openly anti-scientific cultural rela-
tivist/constructivist position which compares science with religion; it
holds the view that there is no truth in scientific knowledge, which
moves within paradigms motivated by social causes;[1] or that the
science of an African tribe's witch doctor is comparable to western
science. I must clarify explicitly that my view of science in general is
realist rather than constructivist/relativist. I may accept some degree
of constructivism in a few speculative areas in science, but in general

[1] Thomas S. Kuhn regrets that many people have misunderstood his work
The Structure of Scientific Revolutions, and terms like "paradigm" (Horgan, 1996,
ch. 2). Not all the crazy ideas generated after the publication of his books
are due to him, but are mostly thanks to those philosophers of science
which took his book as the new bible on which to base the dogma of
relativism and constructivism.

I think that science is talking us about truth in nature, not merely a truth based on social consensus. I should clarify my position because a postmodern philosopher might interpret my criticism of the system of science as an attempt to defend constructivism. One might wonder why, in spite of all the problems that scientific institutions possesses, all the corruption and biases, we should trust that science is talking about truth. My answer is that science in general, all through its history, has shown the robustness of many scientific ideas as solid absolute truths about nature, independent of the social context. There is no doubt from my side that science talks about truths. Nonetheless, science is a slow process and it is quite possible that wrong ideas may dominate science for a long time. My point is that all the problems I have described in the way scientific institutions work nowadays affect the degree of credibility of recent achievements in science, specially the most important ones; but in the long term, I trust that most of the wrong statements will be corrected. Therefore, if somebody asks me whether I believe in science, I will reply: Yes, I do. I love science, I believe in science, and I think I have got good reasons to do so. Is it like a religious belief? I don't think it is, and there are already plenty of books discussing this question, so I will not waste my time, as many philosophers of science do, by repeating again and again the same arguments about it.

Then there are the less crazy trends within the philosophy of science, limited to explaining to the world—in very thick volumes—what a hypothesis is, what induction or deduction are, or the falsifiability of a theory,[2] and such trivialities about the scientific meth-

---

[2] Idea popularized by Karl R. Popper which states that the falsehood of an assertion should be demonstrated by a particular observation or a physical experiment; otherwise we cannot be sure that it is false. Elementary! This and other trivialities fill most of the pages of the works of one of the most distinguished philosophers of science in the twentieth century. Also, some misconceptions such as the confusion of predictability and determinism (see criticism in López Corredoira, 2005, ch. 4), or the multiple errors and misinterpretations about quantum mechanics (see criticism in Peres, 2002). This kind of achievements has served him to receive fourteen honorary doctorates from prestigious universities, which give us an idea of how departments of philosophy in universities are.

Popper said that most philosophers are deeply depressed because they cannot produce anything worthwhile, but not him who says he is the happiest philosopher he knows (interview published at Horgan, 1996, ch. 2). Frankly, I find the situation quite depressing if we have to admit that the

od which are well known to any scientist from their early education, and which do not reveal anything new. They dedicate great efforts to ascertaining what is the meaning of "truth", or how many types of reason exist, or how many types of language analysis or the classification of the different schools with their different "-isms". Sometimes, rather than offering arguments to defend a position, they just use these words (e.g., "You are a scientifist! I am an emergentist!"), and they live with the delusion that mentioning a word is enough to refute or defend a position. Or they dedicate their time to speculating without any basis on the meaning of scientific discoveries beyond what scientists see in their own science, or lending a touch of exotic mystery and confusion to some concepts.

Does this kind of fashion in thoughts in the departments of philosophy of science help science to be better? I agree with the claim by the physicist, Richard Feynman: "Philosophy of science is about as useful to scientists as ornithology is to birds".

Certainly, there is a criticism of science in the discourses of philosophers of science, but it does not help too much because it is usually detached from contact with the real problems which scientists worry about. Professional philosophers make very few attempts to understand the present-day problems of science, though there are some valuable rare exceptions, such as the work by Gillies (2008). In most cases, these problems are only mentioned in order to discredit science in general. Sentences such as "this agrees what we had said ..." and trying to sell some of the metaphysical and paranormal creeds that are the usual merchandise of modern sophists. A Spanish proverb says: "under the Heavens, everybody lives on one's work".

Why is the kind of analysis in the present book so infrequent among the works of professional philosophers? I think there are two main reasons: 1) Most of them do not have knowledge of science at close quarters but only through reading books, which do not reflect the real problems; even in the few cases of philosophers with an education in science to the level of a scientist, they do not dedicate their time to research so they only know about present-day problems by hearsay. 2) They are not interested in revealing the miseries of another profession because they themselves share the same problems in even greater magnitude. With the three points which mark out a

---

best thing a thinker can offer is something like the books by Popper. This decline of western philosophy is not a thing to be very happy about.

twilight of the scientific age, we could draw the same conclusions about professional philosophy:

i) Society is drowning in huge amounts of books with ideas repeated again and again, most of them being about things of little importance for our philosophical *Weltanschauung* (world view). The great conceptions about metaphysics, ontology, aesthetics, ethics were produced long ago. Philosophy in universities is divided nowadays in dozens of specialities without much communication between then, and their specialists produce almost nothing of interest, most of the few important ideas just being old problems with new names.

ii) In the few fields where some important aspects of unsolved questions have arisen, powerful groups control and manipulate the flow of information and push toward a particular ideology. Obstruction of the freedom to initiate a line of research or develop an ideology is more prevalent in the faculty of philosophy than in science. Philosophy congresses are simply imitations of scientific congresses. Censorship of publications is more evident (most being confined to local dissemination rather than international); they have practically no objective criteria and there are no empirical data, so a paper can be rejected whimsically, without even producing an explanation for the rejection. Work positions are nearly always handpicked. Communication with the press, or promotion for the publication of books through editorialising is in the hands of the corresponding *supervedettes*. Propaganda decides the survival of philosophical nonsense, etc.

iii) Individuality is better preserved in philosophy and the humanities than in science. Anyway, the critical mass mechanism of knowledge which arises around the distinguished *supervedettes* (usually people with few if any interesting ideas), the editorial management oriented towards the promotion of general mediocrity, have turned intellectuality into a slave of the economy and popularity rather than an elite of geniuses. In our democratic times, important ideas are identified as those which any man can think about and understand. A superior philosophy, a philosophy for the best men, was lost during the last century, the century of *The Revolt of the Masses* (Ortega y Gasset, 1929). Now, the philosophy which is published and talked about is, with very few exceptions (lost behind piles of rubbish), either pedantic technical questions for academicians or childish popular topics for the masses.

In this panorama, what has the "office-philosopher" to say about science? Nothing different from what happens in his or her own house. At present, professional institutional philosophy has as many problems as institutional science, and reading philosophy of science written by contemporary philosophers working in a university is mostly a waste of time. Therefore, it is not strange that there should be silence about the things criticized along this book. And these problems of science are not going to go away, nor are they to be resolved by any paid philosopher (or sociologist or any group of administrators of culture paid for by the state). These questions are things to be discussed by scientists themselves, and from the inside looking out.

Nonetheless, true "philosophers", in the full sense of the word, even if they do not specialize in scientific questions, may offer us interesting insights about many questions related directly or indirectly with the activity of scientific research. Let us consider in the following pages some significant cases[3].

7.2 Nietzsche

Friedrich Wilhelm Nietzsche (1844-1900) is among the most popular and influential of nineteenth-century philosophers. He was born in Germany (Prussia at that time, indeed), a nationality of which he was not at all proud. His father was a Lutheran pastor, who died when he was four years old. Before he was twenty years old, he showed particular talents in music, literature and languages. In 1864 Nietzsche commenced studies in theology and classical philology. After one semester, and to the anger of his mother, he stopped his theological studies and lost his faith. In the following years, he also began to take a great interest in philosophy, by studying the philosophical works of Schopenhauer and Friedrich Albert Lange. Lange's descriptions of Kant's anti-materialistic philosophy, the rise of European materialism, Europe's increased concern with science, Darwin's theories, and the general rebellion against tradition and authority greatly intrigued Nietzsche. This cultural environment encouraged him to expand his horizons beyond philology and to continue his study of philosophy (WP). He also met the composer Richard Wagner, for whom he would first develop a great affinity and later an enmity. When he was twenty-four years old, he was offered a posi-

---

[3] Note: some paragraphs in this chapter are taken from Wikipedia with very few if any changes with respect to the original source. They will be marked (WP) at their end.

tion as professor of classical philology at the University of Basel. Despite the fact that the offer came at a time when he was considering giving up philology for science, he accepted.

Nietzsche's interest in science is not as well known as his interest in philosophy, philology or music, but science also occupied his mind. During the last years of the 1870s, Nietzsche came under the influence of positivist ideas. As noted in the biography by Andreas-Salomé (1894), Nietzsche planned to dedicate ten years, the 1880s, to studying natural sciences exclusively. He wanted to go the University of Vienna or Paris to study physical sciences and atom theory without writing any book during that time; and after those ten years of retirement and learning he planned to come back and show to the world his new wisdom. He wanted to find the foundations of his ideas in a scientific way. However, things turned out very differently.

In 1879, due to health problems—indeed, he was during most of his life a sick man, with short-sightedness that left him nearly blind, migraine headaches, violent indigestions, syphilis—he had to abandon his position as a philologist in Basel, and nor could he could accomplish his project of a scientific career. Instead he retired alone and concentrated to search for his own truths. He moved to different cities in Switzerland, Italy and France, dedicating his few moments of freedom from migraines to think about philosophy and to write his masterpieces. Instead of looking for truths in science, he looked for them in his personal inspiration, and the 1880s was indeed the most fecund period for his most important works. Because of his aggressive criticisms of many sectors of culture, and his solitary character, Nietzsche gained few friends. Although he tried to flirt with women and even proposed marriage in various occasions, his only female company was his mother and his sister. Towards the end of the 1880s, he became mentally ill and unable to work, and he would remain in a mental hospital until his death in 1900.

We cannot say that Nietzsche has made relevant contributions to the philosophy of science. No, he was not a mediocre specialist uttering trivialities. His philosophy is something much worthier: It constitutes a full vision of our existence, a vision of the meaning of our lives. His philosophy is very rich in ideas and also in literary style, with different possible interpretations. I strongly recommend reading of some of his books. Read them with an open spirit, read them as a poet rather than a logician or a scientist, with the heart, but pay also attention to the great truths contained in Nietzsche's prose. I am not going to offer a summary of the highlights of his thoughts. Within his reflections, most of them expressed in an aphoristic style, there are some ideas which concern the meaning of scientific activity, and that is what I want to bring out here. Following this paragraph, I will provide some quotations from his works, which I will comment on afterwards.

"Men crowd to the light not to see better but to become more brilliant. When we become bright enough, other people will consider us as lights". (*The Wanderer and his Shadow*)

"Why do we try to demonstrate the truth? Because of a broader feeling of power, because of its utility, because it is indispensable. Summing up, in order to gain some advantage. But this is a bias, a signal which indicates that deep down we do not worry about the truth". (*The Will to Power*[4])

Nietzsche is recognized as having great intuition about the psychological processes. It is not by chance that Freud took some inspiration from Nietzsche's ideas. We may observe his intuition of the "will to power" in present-day science. The fight for economic power and social status promotes disagreements among specialist from different fields rather than encouraging them to search for "the Truth" together. Astronomers ask for money because they are disembowelling the cosmos of its secrets; the particle physicists are disembowelling matter of its secrets; the biologist life. What impatient individuals who want to reveal all of nature's mysteries and not leave anything for the next generations! Some data has still not been fully exploited and yet we think about gathering more data. Fast! Before anyone else makes the discovery! Impatience has never been typical of wise people. "I know well your little secret" we could say, it is your will to power. With regard to this topic, Nietzsche has made a deep psychological analysis of men's intentions.

---

[4] Elisabeth Förster-Nietzsche, Nietzsche's sister, compiled *The Will to Power* from Nietzsche's unpublished notebooks, and published it posthumously, taking great liberties with the material. The order in which the different aphorisms were published, classified within some topics and showing some systematization of ideas, is not typical of Nietzsche's free style. Nonetheless, the content of each aphorism is quite close to the ideas of Nietzsche expressed in other works, particularly the emphasis of the "The Will to Power" itself, so I interpret them as a creation of Nietzsche. Anyway, whoever is the author of these aphorisms, their content is worth reading.

The fight among specialists from different branches is reminiscent of the defence of lands in the medieval age. The "authorities of the matter", as they call themselves, are like lords of lands who guard their kingdom fervently. When an intruder tries to insert his nose in a speciality which is not his, he will soon meet a cohort of "authorities" reading him his rights. Generally, the lands are also fenced with a language and symbols, to be crossed only by experts. On some occasions, I would say that formalisms are intended to frighten other people, in order to make entrance difficult.

To say that doing research is collaborating for the peace and fecundity of mankind's progress is rather naive.[5] Nations do not invest in research today because of beautiful phrases. Nations, like people, look for prestige. A country sends its sportsmen to the Olympic Games to win prestige, in order to have people to say: "sportsmen of a certain nationality won a medal ...", and then the national anthem will be played and all that. Next day, the newspapers publish in their pages— "our sportsmen won some medals in ..."—and this "our" makes the reader feel proud to belong to his country and then he will want to produce for his society. In the same way, the state pays scientists, even non-technological ones. If they are not useful for industrial production, they are at least useful for generating prestige. It is very beautiful to find in the news: "a scientist from our research centres has discovered..." It makes citizens believe they live in a real country. There are meetings about science even in undeveloped countries; do they also want to collaborate for peace and the fecundity of mankind's progress while their citizens live in poverty?

From this standpoint, the meaning of scientific research is comparable to the meaning of the attitude of a gorilla who is beating its chest proudly. Of course, we have got our technological applications and the things to make our lives more comfortable. We live longer lives, and have lower child mortality. We have got better weapons to kill our enemies. These advantages give us more power. But to feel proud about science is even beyond these advantages. It is a feeling which is beyond pragmatism. When we look at the starry sky formed by the stars of the Milky Way and we think "I know how the stars are formed and emit their light and heat, I know the position of our solar system in our galaxy...", this gives us a feeling of power over nature, something like conquering lands which previously belonged to the gods, and also a feeling of power over other human beings.

---

[5] This was said by the King of Spain, Juan Carlos I.

We feel ourselves to be like demigods. And, if we think in a pythagorean way, in the sense of believing that our understanding of maths governing the laws of physics is like understanding the whole "Being" of the universe, then our feeling of power is even greater; the character of scientists as new priests or beacons of humanity is more evident.

This gives us an anthropological view on the motivations of science, a pessimistic one. The spirit of the eighteenth century Enlightenment, which supported reason and science as great values of humanity, is shipwrecked with those lucid quotations of Nietzsche's, which claim that knowledge of the truth is merely a means to have power and influence over other human beings. In other words, scientists do not care about nature but about society and their status within it. Certainly, over many centuries we have been living in an illusion, an illusion which was praised by Enlightenment ideas in the eighteenth century but it had been present in the mind of many scientists and philosophers since ancient Greek times. And while the illusion was maintained, people moved without doubting the meaning of that "search for truth" or "wisdom for the sake of wisdom". But Nietzsche has suddenly woken us out of our dreams, and the nihilistic view arises. Indeed, he has not invented a new humanity and he was not pursuing nihilism but trying to surpass nihilism. The will to power was always there, but many people refused to see it in that manner. They preferred to believe in the dogma of a science which looks for truth just because that is beautiful and does honour to humankind.

We might argue against this, that the will to power is a characteristic of human beings in all of their activities, that the humanities, arts, philosophy, etc. are also affected by the same criticism. We might argue that, of course, we cannot avoid people being ambitious but, anyway, there are many ways to exert power, and doing culture is much nobler and acceptable than participating in wars or fighting to get a higher status in other jobs. I do agree with these arguments. I totally agree that the development of science and other cultural expressions is the most noble part of our existence as human beings. However, this is not the point. The point is that the principles which motivate human culture are a farce. In particular, the principles that claim that "knowledge for the sake of knowledge is good" or "knowing truth is good" are questionable. The point is that some people have thought or dreamed that human culture is something separate from nature, and it is not: we are the same naked apes—using the expression coined by Desmond Morris (1967)—that populated the earth thousands of year ago, and with similar instincts. "Knowing the truth" is just a fancy dress for the eternal will of human beings: the will to power.

But science is good, anyway! Yes, science is good, as I have claimed in Chapter 2. It was very good indeed for humanity. It has provided humanity with new insights, opening new doors to wisdom and moving beyond the obscurantism of superstitions, offering us a better life. Wisdom is not the same thing as knowing the truth, but we must admit that knowledge of some truths has made us more learned. But what happens after the major mysteries of nature are solved? After that, scientists are still strongly pushing their science in the name of truth. Will we become more learned, wiser, once we know the infinite tiny details of the many scientific fields and subfields? Certainly, we will not. For instance, we will not become more learned once we know which part of the genome is responsible for the white stripes in an Asian tiger mosquito. Wisdom is another thing; it is not an encyclopaedia. But science is still moving because there are still some "truths" which are unknown, no matter how unimportant these things are. Here is the time when Nietzsche's statements become more important. And this time is now, in the postmodern era after Enlightenment illusion.

Truth for the sake of truth does not make sense. It is a good dark argument to cover the dark instincts of the thirst of power among the administrators of science, with very little scientific spirit. Only truth in the service of wisdom or about important questions for our lives matters. The meaning of scientific research is challenged by the meaning of truth in our lives.

"The calculation of the world, the possibility of expressing with formulae all the things which are happening, is it understanding? What would we understand about a musical composition if we calculated all of the things in it which are calculable and reducible to formulae?" (*The Will to Power*)

"We become thrilled when we think that something is known once its mathematical formulation is gained; it is merely an indication, a description, nothing else…" (*The Will to Power*)

These aphorisms are reflections of the particular way to do science in theoretical physics. Again we find a worry about the meaning of truth, this time in the knowledge of physical laws. Or course, of course, Nietzsche is saying, we know Newton's laws, which are supposed to govern matter. We have got mathematical rules to describe and predict any motion in any phenomenon. We have got the laws which rule gravitation, electromagnetism and light, heat, etc. That is nice. Nonetheless, do not be so naïve as to believe that you know what reality is, what nature is. You have merely described some of the consistencies we can observe in some phenomena, but you have no idea what is going on. Therefore, put aside your arrogance of new priests because your view of scientists has not looked deeply enough at the world. This message connects also with the previous one, that the will to power is the most significant aspect of scientific activity, as well as in many other facets of human culture. Thus, the global message might be something like, you physicists are making the world believe that the truth of existence is in your hands, but your contributions are mere descriptions of minor details that you sell as the great truth because of your thirst of power and influence.

From a purely intellectual point of view, leaving aside the useful engineering applications, the concern of human beings lies in penetrating the understanding of things around us. Religions first tried to reach that understanding through the multiple legends of gods. They failed. Metaphysics had its time among philosophers who tried to have the final word about the world. The time of metaphysicians is over too. Lichtenberg, physicist and philosopher of the eighteenth century, said that thinking in metaphysical terms is like looking for a black cat in a dark room with no cats inside. According to Auguste Comte, French philosopher of the first half of the nineteenth century, there were three ages in the intellectual development of our civilization: the religious one, the metaphysical one, and the scientific one. According to Comte, the first two ages are over, and science has taken its place to offer us answers about the world, what is called "positivism". Nietzsche is going a step beyond Comte's claims and arguing that the three ages are over. Nietzsche's atheist position is well known. He is also usually called "the last metaphysician" because, after his demolishing of metaphysical arguments, metaphysics was no longer taken seriously. And, it seems that we can see an anti-positivist Nietzsche too. Of course, there are still, from the nineteenth to the twenty-first centuries, many people who are living with religion, or with metaphysical ideas, or with positivist ideas, but not him, and not the "Übermensch" (translated as "Overman" or

"Above-Human" or "Superman"). The new intellectual of the highest class, the new aristocrat of thought, has surpassed the three ages, because neither religion nor metaphysics nor science allow man to penetrate deeply in the understanding of our world. If there is a truth which is valuable for our existence in the jungle of our civilization, it is to bear in mind that whatever the truth somebody is offering us there is a will to power in their motivations.

The metaphor of comparing the understanding of nature with the understanding of a musical composition is very sound. We cannot listen to the sound of existence and the harmony of the cosmos once we understand all the physics and astronomy which is available in our libraries. Indeed, rather than using the metaphorical expressions of "sound" or "harmony", we should mention the word which is more related to our worries: "meaning". We cannot catch the meaning of nature or the universe when we contemplate all the collections of laws which mathematically describe it, either in terms of multiple disconnected formulae or with the unification of all laws into one. Again, the use of the word "meaning" is metaphorical, but the meaning of what we are talking about is more direct than the mention of music. Certainly, we have not got a word to describe that thing that we do not possess and we can only refer to it with circumlocutions, so I will continue to use the word "meaning" for it.

The meaning in our lives depends on our sense of nature. Therefore, a science which gives no meaning to nature in our existence is just an erudite amusement. Nonetheless, there are elements of science which surely would have been useful for Nietzsche in the arranging of his philosophical program. For instance, the ideas of the evolution of species and natural selection by Darwin would taught him and us many interesting things on human nature and its will to power.

"I do not understand why we should wish for the complete and absolute power of Truth; (...) from time to time, we should have the chance to take a rest from it, otherwise we will get bored with it". (*The Dawn*)

"Lies are necessary in order to live. This fact forms part of the terrible and enigmatic character of existence. (...) pessimism, that is to say nihilism, takes the value of Truth. But Truth is not the highest value, and even less so the highest power". (*The Will to Power*)

Once more, Nietzsche talks about truth. It is not only the will to power is hidden behind the appearance of the noble search for truth. It is not only that the mathematical formulation of physical laws is merely descriptive without providing a deep understanding of nature. Apart from that, the question is why we should give preference to truth over other values.

This question also affects other areas of culture apart from natural sciences. History, sociology, psychology, etc. are all of them embedded in the search for their truths. In general, even the most primitive tribe has some approach to reality, mainly for its utility. If we live in a jungle, it is useful to distinguish which plants and animals are poisonous and which are not. These are truths which are useful to keep us alive. Nietzsche is not referring to those truths. Neither does he refer to the advancement of medicine or the useful engineering applications of our science. He refers truth in its purest sense, truth for the sake of truth and nothing else.

His position has been confused many times with the relativist position of post-modern philosophers, who deny any truth available to mankind, and claim that all knowledge is a social construct. No, in my opinion, Nietzsche's position is not that. Indeed, he states clearly that there is some truth which is already in our hands: nihilism (of moral values), pessimism, an absence of freedom of the will, as it is expressed by Nietzsche in many works, a fatalism implicit in the almighty nature. He would even claim the "Amor Fati" (the love of fate) in his texts, sharing with Spinoza the joy of contemplating a nature for which we are puppets. Nonetheless, Nietzsche also claims that we must escape from that truth. Lies are necessary in order to live, maybe in the form of poetry, art, music, maybe in the form of some naïve belief, including the illusion that we are able to subdue our fate when we make our best effort. Because, life is the highest value of our existence, and if lies are necessary in order to live, we must sometimes abandon truth. Not always, only sometimes, in order to avoid life being drowned of with so much boring science.

There is in Nietzsche a spirit of liberation, which must not be confused with a devil-may-care attitude. He was a learned man, with a knowledge of many languages, classical culture, philosophy, music, etc. not a lazy student who does not want to study physics because it is boring. His intellectual discourse was to overtake the old morality which claimed that certain kinds of slavery are good for man. Why should we be condemned to study or to do research on a topic which doesn't excite for us? Certainly, at the end of the nineteenth century most advances in science were enthusiastic adventures, but there was already an increasing load of knowledge which served only to fill the encyclopaedia with boring data. At present, most if not all research is of that kind. Why should we work hard, dedicating a lot of time and money (which also means the efforts of other people in society) in order to make a tiny advance in our global understanding of truth? Is it because of the value of truth? But truth is not the highest value.

Certainly, many scientists nowadays are unaware of Nietzsche's views on truth and science. Better for them, as they will keep the flame of their illusion for a longer while. Ignorance is sometimes useful. Nevertheless, Nietzsche's ideas are deeply embedded in our society. Nietzsche was perhaps the philosopher with the greatest influence on twentieth-century thought. Therefore, sooner or later, the result of his work will reach science, and scientists some day will also wonder whether truth is the highest value. And if scientists do not, then society will remind them to do so.

7.3 Miguel de Unamuno

Miguel de Unamuno y Jugo (1864-1936) was a Spanish philosopher and writer. He worked in all the major literary genres: essays, novels, poetry, and theatre, and did much to dissolve the boundaries between genres and create new genres. He was one of the foremost representatives of the "Generation '98" movement in Spain, a generation of intellectuals associated with the epoch of the Spanish crisis of 1898 after the loss of the last Spanish colonies in America and Asia: Cuba and Philippines. In 1880 he entered the University of Madrid, where he studied philosophy and arts, receiving his Ph.D. four years later. His dissertation dealt with the origin and prehistory of his Basque ancestors. His early years were deeply religious, but in Madrid he started to visit the "Ateneo", sometimes called the centre of blasphemy in the city. In its library, he read the works of liberal writers, and he would move closer to positivist ideas. After completing his doctorate, Unamuno worked as a private tutor in Bilbao, where he also founded, with his friends, the socialist journal *La Lucha de Clases*. From Bilbao he moved to Salamanca, to assume the chair of Greek at the university. In 1891 he married Concepción Lizárraga Ecénnarro; they would have ten children.

During a certain night in March, 1897, he went through an existentialist crisis. He sensed the loss of existence after death, he sensed the nothingness. His wife tried to console him over it but without too much success. After this experience, Unamuno would change radically his view. From universal philosophical constructions and outer reality, he turned his attention to the individual person, and to their inner spiritual struggles in the face of questions of death and immortality. Unamuno once stated: "Wisdom is to science what death is to life or, if you want, wisdom is to death what science is to life". His greatest works would be inspired by that existentialist anguish. His positivism and his faith in science and progress would collapse. He concluded that the anguish of mankind is due to the longing for immortality. Seeing reason leads to despair, Unamuno concluded that one must abandon all pretence of rationalism and embrace faith, stating that all religions have some truth in the sense that they serve as consolation for the fact of being born only to die some day. Indeed, his rationalist position is atheist rather than theist, but he defended the use of religion from a sentimental point of view.

In 1901 Unamuno became rector of the University of Salamanca; he held the post intermittently until his death. He was relieved of his duties for the first time in 1914, for political reasons. In 1924 he was exiled to Fuerteventura in the Canary Islands for opposing the military dictatorship of general Primo de Rivera. After a few months, he escaped to Paris, where his friends helped him draw international attention to his exile. He then settled in Hendaye, the French Basque town nearest to the Spanish frontier, where he spent five years. General Rivera died in 1930 and Unamuno went back to the University of Salamanca, and was re-elected rector in 1931. He worked as professor of the history of the Spanish language, and continued to criticize political parties of both the left and the right, but in 1936 he was removed once again, this time for denouncing Francisco Franco's falangism. Unamuno was placed under house arrest. He died in Salamanca on the last day of 1936, a few months after the outbreak of the Spanish Civil War.

Unamuno was a highly cultured humanist; he could read in fourteen languages, and he knew about most of the intellectual developments of his time. He was, like Nietzsche, a philosopher without a rational system. Rather, he oriented himself towards an irrational philosophy. He objected strongly to academic philosophers. As noted, he stressed that the deepest of all human desires is the hunger for personal immortality despite our rational knowledge of life. He also railed against the fashion for embracing cultural values from dominant central and north European countries (Germany, France,

UK) in Spain; rather, he defended the expansion of Spanish values into other countries. His literary and philosophical works cover a wide spectrum of interests. Here I will focus on the aspects related to science.

Among his most important pure philosophical works is *The Tragic Sense of Life in Men and in Peoples* (Del sentimiento trágico de la vida en los hombres y en los pueblos, 1913). There are also many philosophical reflections in his novels. For instance, his last novel *Saint Manuel Bueno, Martyr* (1930) focuses on a country catholic priest, Don Manuel Bueno, who does not believe in the God and the eternal life that he preaches. However, Don Manuel continues to take care of his parishioners, revealing his tragic secret only to a few people before his death. Why preach something that we know is false? "Truth? Lázaro, truth is by any measure something terrible, something intolerable, something mortal. The common people could not live with it" (Unamuno, 1930).

In common with the emergence of an irrational philosophy in Europe, of which Nietzsche was an outstanding representative, Unamuno shares the view that truth is not always a good thing, and might even be harmful for life. He was not one of those postmodern relativists who denies the existence of an absolute truth, nor one of those who claim that our science cannot reach any truth. No, he was not. Rather, he was a well-informed thinker who tried to understand what really gives meaning to the life of human beings. And he concluded that cold truths ascertained by science, the image of the human being shown by biology in which we are simple animals with an ephemeral life, do not offer that much consolation. We feel the tragedy of our limited existence, and science is unable to offer us a solution for the most important of our worries.

Possibly, Unamuno was not against the development of science. I interpret his work as suggesting that he was not insinuating that superstitions and religions should be preserved and science should vanish. No, he was an intellectual man who was aware of the importance for our culture of the development of science until the end of the nineteenth century. He was not a nostalgic metaphysician who wanted to go back into history. Nonetheless, he realized that at the moment in which he was living, science was losing its way and walking towards a fate—the absolute knowledge of everything—which was, apart from being unachievable, undesirable too. We could also consider this, with more reason now: We have classified millions of species of animals and plants, we have catalogued millions of galaxies, we have decoded large portions of the genes of many species … what can we do with all that? I mean, we know that many of the things that we know through science are part of the truth, and in some senses, it is nice to have so much knowledge. But, once we have established the most important elements of physics, biology, geology, chemistry, etc., useful for our lives and for the control of the nature around us, why should we dive deeper and deeper in the bottomless well of truth? When we suffer an existentialist crisis, are we going to sleep better, if we keep an encyclopaedia on our bedside table, in which we can consult at midnight which part of the genome is responsible for the white stripes in an Asian tiger mosquito? Possibly not, since our worries as human beings are not directly related to our knowledge. Of course, we are more than animals who worry about food, sex, power in the tribe… Let me correct that: "some of us" are more than animals (but we are a minority, because most common people just worry about food, sex, power…) and we pursue some sort of "culture", a culture which makes us superior to animals, more spiritual (in a non-religious sense), more elevated. Is science part of this culture? Definitively yes, although a part of culture which cannot substitute for other parts.

For Unamuno, it was very good that some countries dedicated time and effort to developing science and technology. Germany, France, Great Britain and a few other countries developed science enormously during the nineteenth century, and that was good for humanity in general. However, Unamuno was not in favour of the expansion of this positivist spirit throughout the world. Science was just one of the possible manifestations of culture in human beings, but there were many other. In particular, he was reluctant to make Spain a land of scientists and to push towards a politics of promoting science in this country. He kept up a dispute with the Spanish philosopher José Ortega y Gasset, who was in favour of moving Spain closer to events in the rest of Europe, and he argued in a letter to him:

> "Do they invent things? Invent them! Electric light is here as good as in the place where it was invented. [...] On the one hand, science with its applications is useful for making life easier. On the other hand, it is useful to open a new door to the wisdom. And are there no other doors? Have not we another one?"

"Spain is different" was a tourist slogan in the sixties, and certainly it is, or it was. Now the unique character of each nation is being diluted. What a pity! What Unamuno claimed is that Spain, or any other country, is not Germany, that our culture is different. Each society has different ways of expressing higher ways of existence. The character of people is not the same everywhere. And the reasons which motivated some individuals to create science in some countries were not universal. At present, we live in an epoch in which even developing countries, with traditions very far from central European perspectives, are dedicating great efforts to producing science. For Unamuno, that would be an error. India, for instance, has other ways of expressing its wisdom, different from producing western-style science. And in some cases, it is as shocking that some countries try to produce science as it would be to see a Japanese man playing flamenco music. Of course, it is possible, and indeed some oriental people are very good at playing Spanish music, but this is because they have lost partly the spirit of their nations, becoming part of a globalized culture, dominated by a western perspective. Whether this is good or not is a topic for a long debate. For the moment, it is worth noting that the dominant countries in science continue to be those with a long tradition of it, although soon we will see how China or similar countries produce leading scientists, pianists, etc. The United States does not have a longstanding tradition, prior to the twentieth century, but their tradition draws on their Anglo-Saxon culture. And Spain, although it has made important progresses in its application to science, and now dedicates an important amount of money to it, comparable to other European countries, does not have the same productivity in science as yet. Maybe competition with Germans in the innovation of science and technology is not yet part of the Spanish spirit, in spite of the globalization of culture.

The expressions "invent them!" or "let them invent!" is typical of Unamuno when asked whether Spain should emulate other European countries in science and technology. It may look somewhat cynical or simply a rebuff, but it is more than that. Unamuno thought that we should leave the investment in science to other nations, that their own pride would announce the news to the world and the interesting ideas would arrive in every country. Science is good, but it is universal; knowledge in pure science is not the private property of a nation. Of course, applied science in technology is another matter, as a country may receive royalties for its innovations.

But, apart from technology, in the relationship of to wisdom, science is universal and it has no owner. In the mentality of some nations, the pursuit of scientific discoveries is a part of the way they express their culture, in the same way that Indian songs are another manifestation of a culture. But why should every nation create Indian songs or attempt the unification of physics? There is no reason for that.

Learned people should pay attention to all kinds of wisdom in different cultures, and science is one of them. Certainly, we may say that science is more than an expression of a culture in the sense that it contains "absolute" truths, which are valid everywhere. But, the thing is, worrying about finding an absolute truth of nature is not universal. That is, although scientific statements are valid everywhere, the value of science is not the same everywhere. We could say that our science is right and other world views are wrong; yes, we can say that. But we cannot say that, because of it, we have the right to impose our culture of science and technology on other civilizations, different from western ones. No civilization is right or wrong in that sense. From an anthropological/sociological point of view, culture is just the way that a civilization chooses to express its existence.

Human beings are moved by irrational fears and wishes. The concern about truth, the interest in logic and reason and science, is just a peculiarity of a culture in a particular place and a time. It is not a universal characteristic of human beings. What is universal is our fear of death, and there have been many ways bearing this anguish: through religion, or through those philosophies which talk us about the wisdom of how to die with dignity and without fear (for example, the death of Socrates), or through trust in medicine and high-tech devices which may make our lives longer, etc. In this respect, embalming and burying the ancient Egyptian pharaohs under pyramids in the desert makes no less sense than installing an artificial heart to prolong life in someone who is dying. We live with illusions in order to cheat our fears. At present, we live with the illusion that a longer life with a lot of medicine and surgical operations is better than a shorter life without visiting physicians. However, we are going to die, and there is no life after death; that is the most important truth, for which science cannot produce anything more.

Possibly, the clearest thought by Unamuno about science can be found in the quotation that I gave at the beginning of this book, which I repeat here:

"Yes, yes, I see it; a huge social activity, a powerful civilization, a lot of science, a lot of art, a lot of industry, a lot of morality, and then, when we have filled the world with industrial wonders, with large factories, with paths, with museums, with libraries, we will fall down exhausted near all this, and it will be for
whom? Was man made for science or science made for man?"
(Miguel de Unamuno – *Tragic Sense of Life in Men and in Peoples*)

Unamuno claims here that we have already got too much science, too much art, too many history books, too many academicians,
too many libraries ... and this was a century ago. Now the situation
is much more astonishing. Certainly, we are a powerful civilization.
As a whole, the quantity of information and ideas we have collected
is impressive. But, individually, what are the benefits for a particular
person of this vast culture, which cannot be embraced even in a
small amount by any one human being? In the seventeenth and
eighteenth centuries, a man could still enjoy all the varieties of culture: science, arts, humanities, etc. But what is the point of expand-
ing the limits of our knowledge more and more if one life is not
enough to digest even a small part of all it at present?

The same thing happens with modern computer technology. I
know people who are very proud because they have got computer
hard disks with many gigabytes of memory, and they can save tens of
thousands of songs to listen to. But this is absurd, I think, because
you will never have time to listen to tens of thousands of songs.

There is an obsession that human beings have with making
things bigger than necessary. The spirit which accompanied the
builders of the pyramids in Egypt or the cathedrals in the Middle
Age is behind the creators of huge telescopes or particle accelerators.
But a man cannot reach heaven through the building of huge archi-
tectural structures, and a man cannot reach the whole truth by
accumulating vast amounts of scientific knowledge. And by building
these mega-structures, we are caught in a kind of slavery. "Was man
made for science or science made for man?" Apparently, man was
made to work obsessively for his delusions of grandeur. Indeed, in
this process of civilization, some people have these delusions, while
other people provide the manpower to do the routine tasks.

For whom? Perhaps, the historical moment when we must once
again raise the question has already arrived. Where are we going? The
scientific method is worn out. Are not we contemplating the twilight
of the scientific age?

7.4 Oswald Spengler: The Decline of the West

Talking about culture as part of the process in the development of civilizations, I think an important name to mention is that of Oswald Manuel Arnold Gottfried Spengler (1880-1936), a many-sided thinker. He was born in Germany in 1880. In secondary school, he became interested in many disciplines. After his father's death in 1901, he attended several German universities as a private scholar, taking courses in a wide range of subjects: history, philosophy, mathematics, natural science, literature, classical culture, music, and fine arts. His private studies were undirected. In 1903, he failed to earn his Ph.D. with a thesis on Heraclitus because of insufficient references in his work (WP). He would earn the title in 1904, but found few opportunities to continue in an academic career. We could say that Spengler suffered the consequences of trying to be a learned man, in the style of Leonardo da Vinci or Goethe, rather than being a specialized academician who included many references in his texts. Already at the beginning of the twentieth century, the academy clearly manifested a preference for technician/specialists and rejected the attempts of giants of culture who tried to dominate vast parts of it. Perhaps, Spengler was one the last great men who held a broad philosophical overview, integrating the whole of culture. Nowadays, universities are populated by dwarves specializing in small fields, including the faculties of philosophy.

In 1905, Spengler suffered a nervous breakdown. As in the case of Nietzsche, Spengler's health was not good, and he suffered several health problems, including migraine headaches and anxiety, throughout his life. Nietzsche, by the way, would have been an important influence in Spengler since his early education in the secondary school. In the years until 1911, he would work as a teacher of mathematics, physics, history and German literature in several secondary schools. His mother died in 1911, leaving him a modest inheritance, so he decided to leave the teaching profession and move to Munich, where he would live until his death, to dedicate his time for his intellectual activities. He was living alone as a cloistered scholar, doing a few jobs as tutor or writing for magazines to earn some additional income. During World War I, he was not called up for military service due to his congenital heart problem. In that period, he suffered poverty since his inheritance had been invested overseas.

His main creation is the magnum opus *The Decline of the West (Der Untergang des Abendlandes)*, a treatise covering most of the fields of our culture from a historical point of view, comparing our civilization with others in the past, and arguing that we have entered in an epoch of decadence for the western culture. He began this work before World War I, inspired perhaps by Otto Seeck's work, *The Decline of Antiquity*, and by the events that he was observing in Europe at that time. He finished its first part in 1914, delaying its publication till 1918 because of the war. When it came out, it became a success. It comforted Germans because it seemingly rationalized their downfall as part of a larger world processes. It is not important here whether it was successful among the masses; the number of copies sold is not important. It is remarkable that it also became a success among many talented intellectuals of the epoch. The book met with wide success outside Germany as well, and by 1919 had been translated into several other languages. Many professional academicians (historians, philosophers, sociologists, etc.) criticized the book, saying that it contained several errors, and claiming that it was the work of a dilettante. In my opinion, most of these academicians became irritated because it showed that an outsider could create such a magnificent work, something which they in their academies were unable to create. Certainly, it might contain some errors in particular subjects, something for which a specialized dwarf could find reason to criticize, as well as some missing references; however, dwarves cannot criticize the broader work of a giant. Spengler rejected a subsequent offer to become Professor of Philosophy at the University of Göttingen, saying he needed time to focus on writing. In 1922, Spengler issued a revised edition of the first volume containing minor corrections and revisions, and the year after saw the appearance of the second volume.

In the following years, he would write some other minor works, he would try getting into politics without success, and finally he would die of a heart attack in 1936. Although highly influential and internationally popular during the interwar period, Oswald Spengler's work fell into intellectual disrepute and obscurity following World War II. This has to do possibly with the politically incorrect ideas that he expressed against the Weimar Republic in Germany, and against democracy in general. He was not a Nazi and never took to them.[6] However, the new tastes of the intellectuals sheltered by the

---

[6] There are people who think that Spengler was a promoter of the German *Nationalsozialismus* who reached the power in Germany in 1933 with Hitler. That is wrong. Spengler was never a member of the NSDAP, and indeed he

dominant Anglo-Saxon culture after World War II were not very friendly towards Spengler's works. One example can be found in *A History of Western Philosophy* by the British philosopher Bertrand Russell, who claimed that Spengler was ignorant of history, and distorted the facts.[7] In spite of it, he is still an author who is widely read, with many admirers among learned people.

We come again to his work, *The Decline of the West*. I think its force resides in its attempt to understand the engines of history without falling in the metaphysical idealist speculations of other previous philosophers, like Hegel. I do not agree all of Spengler's statements, but I consider his book to be a masterpiece in many ways, a very recommendable book, with plenty of lucid ideas and admirable global vision; a giant, a brave thinker with a strong character and something interesting to tell, rather than a boring treatise of trivialities and diplomatic sentences of the kind so common among our dwarf-philosophers.

Spengler tries to express the idea of the universe as nature—a mechanical thing, a computable thing, a set of everything which is necessary according to eternal laws (and he thinks about the universe as history), of fate, of experience of lived life and without relationship with maths, of a truth which is constantly changing. His idea of fate can only be communicated through the arts and, precisely because of that he thinks that the arts are very important in culture. Within this conception of the world, terms like "nation", "origin", "fate" make sense, whereas they cannot be understood from a rational cosmopolitan perspective. The people and the state are

---

did not have a good relationship with that party. Gregor Strasser and Ernst Hanfstängl tried to recruit him without success. He voted for Hitler and against Paul von Hindenburg in 1932, like many Germans, but once Hitler came to power, he was against his politics. Spengler had an interview with Hitler only once, in July of 1933, but they found no affinity with one another, and there was no collaboration. The Nazis would also ban some works by Spengler in which Nationalsozialismus was criticized.

[7] In my opinion, Russell is a mediocre academician quite biased about history, who is not in a position to give lessons about history to a giant of that subject like Spengler. It is enough to read the chapter about Romanticism in his book in his *History of Western Philosophy* to realize how he puts Great Britain at the centre of the world and neglects important authors of continental European philosophy; including authors like Jane Austen, Mary Shelley, D. H. Lawrence among the important thinkers, or even Lord Byron as one of the eight most important philosophers of the nineteenth century.

forces of human existence. The motor of history resides in the struggles of these nations; they are nothing to do with ideas on moral, truth or justice. The true history is the one produced by the men of action, like Alexander the Great or Napoleon Bonaparte, rather than the politics discussed by theoreticians, the history of the instincts of possession, of the will to power and the eagerness of pillage, the history of facts, of living peoples. In this world, truths are those which work effectively in history. Spengler mocks the concepts of freedom, human rights, humanity, progress, justice, virtue, truth, reason, finality and the great abstract theories about status or politics in general. Rights are the expression of the force of the dominant class. Money is the true power in the democratic societies rather than the constitution or any law.

This humanistic view, in which all cultural products are just the fruit of an epoch, a vision that I agree with, leads him towards a pronounced relativism or instrumentalism, by which all truth in science is temporary, a view that I do not share and have classified as postmodern stupidity in other places in this book. In this way, Spengler shares the views of many dull philosophers, who mix all the products of culture (arts, humanities and sciences) and claim that everything is a cultural product without any "natural" truth in it. According to him, there are as many kinds of maths as there are cultures, and each culture or civilization has its own way of seeing nature; the prejudices determine the scientific knowledge. There is no difference between a catholic view and a materialist view of the universe, he says; the deep religiosity has the same tone as the atheist formulation of modern dynamics. Both of them claim a religious necessity and a way to know God. For Spengler, physics is a matter of faith, and atomic theory a myth. According to him, the raising of a scientific problem implicitly carries with it the corresponding solution. Apparently, he does not take into account the empirical aspect of science and the comparison of ideas with the facts. He only sees the human part in the development of scientific ideas. This is in my opinion the most unfortunate part of *The Decline of the West*, although I appreciate that his level of erudition and knowledge about the sciences shows his arguments are at a much higher level here than among most philosophers involved in the same discourse. Certainly, Spengler's opinions about the contents of natural sciences are easy to forget. Nonetheless, in the human/social aspects of science, as well as in other manifestations of culture, I think there is something worth paying attention to.

Spengler's major conclusions are twofold:

1) There are three possible ways of a civilization developing: within magical, Apollonian or Faustian conceptions of the world. Magical conceptions are associated with primitive cultures dominated by animism, religions and other superstitions, of which there are still some remnants in present-day societies. The Apollonian spirit is the characteristic of reason and logics, of the quantification of nature, typical of most scientific activities. The Faustian spirit is characterized by the will, the feeling of a direction towards an end, the consciousness of history.

2) The second conclusion is that the history of our culture can be compared with the history of other civilizations and their developments, and in making that comparison it can be observed that western culture is declining, it is reaching the end of its possibilities. The epoch of the arts is over, and now only rubbish is being produced, as an illusion of art, keeping it alive. The epoch of philosophy is over. And the epoch of science is over too.

"Death of science consists of the existence of nobody able to live it. But 200 years of scientific orgies get fed up in the end. It is not the individual but the spirit of a culture who gets fed up. And this is manifest by sending to the historical world of nowadays researchers who are more and more small, mean, narrow and infecund". (Oswald Spengler, *The Decline of the West*)

For Spengler, the epoch of mathematics is over, and there now remains only the work of conservation, refining, polishing, and selection. Physics is also near its limit. What remains now is an industrial production of hypotheses. Still, in the 1910s and 1920s, when Spengler wrote about this, there were important discoveries being made in physics. Nonetheless, I think Spengler is right in his prediction about the decline of the scientific world, as well as the decline of culture in general, a trend which may last many decades or centuries. In my opinion, he was ahead of his time, and he saw the problems of the future in our civilization in a prophetic way. We must bear in mind that the decadence of the Roman Empire lasted almost three centuries, after the death of Marcus Aurelius, with certain fluctuations but following a general average trend of decline (Gibbon, 1776-1789). The same thing may occur with the decline of our culture now: We have been declining all through the twentieth century, and we will continue to do it during the twenty-first century.

These thoughts in some way reflect the pessimism generated in Europe during and after World War I, with the downfall of the great empires. The Austrian writer, Stefan Zweig, in his autobiography *The World of Yesterday* talks nostalgically about the great epoch for central Europe between 1884 and 1914, an epoch of splendour in which art and culture flourished among educated people, and idealism and high values among many young people, of a general liking for poetry, theatre, etc. And how all that was quickly devastated afterwards in an ill-fated environment of battles led by men without talent, in a Europe which had lost trust in most values except money. Certainly, Spengler understood perfectly that the problem of our epoch is the identification of money with the strongest power in democratic societies. Money chooses the dominant democratic parties because the mass media is under its control. Our present-day society lives in that state of decadence, which is not something new because other civilizations in the past have passed through similar circumstances. This period of decadence and capitalism (the dictatorship of money) will continue to drag down our societies until some radical change of order can be produced, which he compares with the arrival of Julius Caesar in the Roman Empire, to replace the decadent Roman republic preceding it.

There were also other prophets of the end of history, both before Spengler—the cases of Hegel or Marx—and later, even quite recently—Alexandre Kojève, Arnold J Toynbee or Francis Fukuyama, for instance. In my opinion, none of them approach the topic in the best way and they are mostly entangled in optimistic and naïve opinions, by which we have arrived, or we have almost arrived, at the best of all possible worlds, so no further historical evolution of the political idea is necessary. Arnold J Toynbee perhaps holds an intermediate position between the optimistic and pessimistic positions: In his *A Study of History*, he rejected Spengler's fatalistic view that civilizations rise and fall according to a natural and inevitable cycle, proposing that a civilization might or might not continue to thrive—"Civilizations die from suicide, not by murder"—, depending on the challenges it faced and its responses to them.

Among the extremely naïve thinkers, we have the present-day US academician of political sciences Francis Fukuyama (1989, 1992), who argues that liberal democracy may constitute the end point of mankind's ideological evolution and the final form of human government, and as such constitutes the end of history. His work, *The End of History and the Last Man* (1992) was a bestseller, and it is a good representation of the present-day political thinking in the United States. There is no doubt that he is much more politically correct than Spengler.

How can an expert in politics believe that there is one form of government which will last forever? Is Fukuyama joking? No, he is not joking, and apparently he is very convinced about what he says. He says, for instance, that there are no barbarians at the gates who might produce a revolution in the present-day system. About ecological or similar problems, he thinks the solution will come with the development of alternative technologies, or technologies intended to protect the environment. After reading this book and seeing that it is one of those most cited about the topic, even more so than Spengler, I have no doubt that our civilization has entered deeply in its decline.

Fukuyama (1992, ch. 6) wonders whether scientific activity may cease: "Can the scientific method cease to dominate our lives, and is it possible for industrialized societies to return to pre-modern, prescientific ones? Is the directionality of history, in short, reversible?" He answers these questions in chapter 7 of his book:

> "…the ability to use modern natural science for military purposes will continue to give such states advantages over states that do not. [...] Man´s post-cataclysmic dependence on modern natural science would be even greater if it were ecological in nature, since technology might be the only way of making the earth habitable once again. [...] And if the grip of a progressive modern natural science is irreversible, then a directional history and all of the other variegated economic, social, and political consequences that flow from it are also not reversible in any fundamental sense".

At this point, and of interest in our discussion on the end of science, we have to distinguish between the development of technology or applied science, which, as claimed by Fukuyama, will offer some advantages to society, and the development of pure sciences. Fukuyama mixes everything together and he contends that no change in research policy will occur in the future. Because of the reasons I have already expressed through this book, I do not share his opinion with regard to pure sciences. I may agree that the cumulative knowledge of science will have a long life, although it is also possible that a new Dark Age will come in, in which many advances of science will be forgotten. In any case, thinking that mankind will always invest efforts in doing research because this is an irreversible process of civilizations is not a good position, in my opinion.

The decline of the West (or better to say "twilight", since the German word *Untergang* means twilight when applied to the sun) is a conclusion derived from patient observations of culture as a whole rather than a mere frivolity. It is a conception of the culture as being comparable to an organism, in which there is a birth, growth, reproduction and death, all of them with some symptoms which are recognizable. Nobody likes to be called moribund, or the representatives of decadence in a culture. Artists, scientists, humanists, and also politicians, economists, and leaders of the society, like to dream about their delusions of grandeur, drawing on western culture from earlier centuries. Our artists like to be called the contemporary Michelangelo or Beethoven. Our scientists want to be called the contemporary Galileo or Darwin. Our philosophers want to be compared with Plato or Kant. And the same thing occurs for other groups. But the fact is that such comparisons are not possible. The classical authors are several orders of magnitude more important than the best of our contemporary authors. It is not only a question of capability, since there are contemporary authors who are very good at what they do, but they are not in the appropriate place and the appropriate epoch. For science, there are brains as well prepared as Einstein's brain or even better, but it is not the moment to reinvent general relativity. Culture as a whole behaves like an organism, and individuals cannot make a moribund culture last longer because it is beyond their capability. Our contemporary artists, scientists, humanists, politicians, etc. try to sustain the delusion of a golden age for our culture, or that we live in the best political situation, through propaganda such as Fukuyama's book. And they try to bury lucid classical works, such as Spengler's, because they consider it too pessimistic and they do not like to see things as they are.

In contrast to the idea of the dawn and twilight in a culture, there is also the extended idea that all epochs are equally important, that there is no rise or fall of a civilization, that our epoch is as important as the previous ones, and the people who are talking about a better epoch in the past are just nostalgic people who cannot adapt themselves to the present times. This idea is illustrated, for instance, in the recent film written and directed by Woody Allen, *Midnight in Paris* (2011), which concludes that there are no golden epochs, and that there are always people who think the past was better, but this is just a psychological mechanism which does not offer an objective representation of reality. This is a relativist idea with which I do not agree. Certainly, there may be many biases in the evaluation of which epoch was better, and this depends of course on the topic. But it is certain too that decadence is a fact in some epochs of some civilizations, beyond subjective impressions. Most likely, there were in the middle of the fifth century AD in Rome people who thought that their empire had been better in various epochs of previous centuries, both politically and culturally, and they were right: The decline of the Roman Empire and classical culture was a fact. By the same token, it is not a caprice to call that glorious dawn of the classical culture in the fifteenth century, the Renaissance. In the Woody Allen film, a contemporary writer thought that the best epoch for the arts in Paris was the 1920s, whereas for people living in the 1920s the golden epoch of Parisian art was in the second half of the nineteenth century, whereas for the artists living in the second half of the nineteenth century the golden epoch was beforehand, and so on, so on. This means that there will always be some nostalgic people looking at the past for a golden epoch, but it also means, and this is my interpretation, that the glory of the arts has being declining for more than a century. This pessimistic view of our times with regard to science is what I wanted to discuss here.

Talking about the contents of *The Twilight of the Scientific Age*, Lee Smolin told me: "One piece of advice at least for the English speaking world is frame things more positively. Better to speak of dawn than twilight". What a pity that the current dominant Anglo-Saxon culture is so childish as to prefer a sweet lie rather than a bitter truth. History will put everybody in their place, and I guess that within fifty or hundred years very few people if any will remember Fukuyama's awkward statements whereas Spengler's work will be included in the list of immortal authors, provided that there is somebody at that time who still reads books.

8   NONSENSE AND THE SEARCH FOR A NEW HUMANITY

Science is becoming a nonsense for humanity. During the last century, science has advanced more and more in technical terms, more and more in its investment in very expensive experiments, in the amount of information it generates, but it has gone backwards with regard to its motivation. The force which pushed humanity to walk towards knowledge, enlightenment and reason is now pushing very weakly. Now, science continues to work because of its inertia but is subject to some friction because to its erosion. Our science is tired, exhausted. It walks entangled with economic forces rather than with human dreams. Science has lost its first attractiveness; only simple technical operations remain. What is the thing in whose name we do research? Truth? Economy? Prestige? Already at the beginning of the twentieth century, Max Weber thought that the dreams of science as a way to the truth, or happiness, or knowledge of God, etc. were shipwrecked. Neither is the scientist a prophet, says Weber. Science as an amusement still remains but the growing pedantry and smugness limits it. Dreams remain perhaps in the mind of some amateur scientists, who try to imitate the exploits of the geniuses of history of science, but these imitations assume a ridiculous position, like a Quixote living on the fantasies of a past epoch.

Our science has become an animal without a soul, or it might be better to say, a colony of animals, a group of organisms which devour human efforts and do not offer anything but growth for the sake of growth. Scientific organizations behave like a colony of bacteria which reproduce as far as the available food and money allow. The more you feed them, the more they grows: more Ph.D. students, postdocs, staff researchers, supercomputers, telescopes, particle accelerators, papers, etc. And, if the money tap is closed, the people dedicated to science and their by-products are proportionally reduced. This is not the science of Galileo, Darwin, or Einstein, who produced their ideas when they felt the "spiritual" necessity to express them, independently of whether they were paid for that or not; certainly for prestige, and for the pride in revealing new truths. Science, like philosophy or the humanities, like the arts (except architecture and other fields which were subject to economic funding) was produced in our civilization as an expression of the soul of our western culture, not necessarily associated with the status quo. Nowadays, there are very few things to express; almost everything in science is reduced to find a small fiefdom of nature to analyse— whether there is any fundamental question to solve in this analysis does not matter—, and publishing papers on it and getting citations from colleagues with the aim of getting jobs and extra money for expenses. Getting money to employ more Ph.D. students, postdocs … and when these students and postdocs grow up, they become new senior researchers who ask more money, and so on. The sense of all this industry is one of primitive life: just a struggle for survival and spreading genes.

We have created a new organism, a new monster: an autonomous structure of knowledge. And the reason for keeping alive this new organism is the same as the meaning of life for any primitive being. But why keep alive autonomous structures which are not part of ourselves. This enterprise would have a meaning for us if it gave some sense to our lives, but that meaning was lost on the road. Why should science survive if it does not give better lives to the human beings who create it? It was important for our understanding of nature in the past, but it is not so much now. Our philosophy of nature does not change, our *Weltanschauung* (world view) does not change with the latest discoveries; only subtle details are now produced. Is it good for the individuals of the mankind to know so many details about nature? And if it is not produced for each of us, "it will be, for whom?" (Unamuno).

Unamuno, yes, Unamuno is mentioned again in the final chapter of this book, a variation on a theme drawn from him. "For whom? For whom?" should sound like a hypnotic mantra which pulls us into a deep meditation on the human condition. And connected with this "For whom?", one of the main topics in Unamuno's philosophy: death. Do you not see that all the subtle and vast knowledge you are building cannot be assimilated by any individual in anything less than many thousands of years, and our lives are much shorter than that? Even science cannot console you for the fact you will have to die, Unamuno might say. We could reply that it will serve for the generations after us. And after the death of our civilization? Luminet (2008) argued that man's achievement must inevitably be buried beneath the debris of a universe in ruins, but man's efforts are still noble because he knows that he will die, and the universe knows nothing of this. I agree that the efforts of scientists are noble, and I agree that one of the main distinctions between human intelligence and inert matter is that we do realize what we are and our finite nature; we have understood our nature thanks to our science, and that is great. No so great are the millions of boring papers containing unimportant details, and these will also be buried beneath the debris of our civilization without having given any meaning to our existence.

Contemporary science has become Sisyphean, that is, as endless and unavailing as labour or a task. The Greek myth of Sisyphus tells us how a king was punished by being compelled to roll an immense boulder up a hill, only to watch it roll back down, and to repeat this throughout eternity. Are we also punished by endlessly doing science?

Indeed, life itself is Sisyphean. Those passions of human beings, all that suffering and tears and illusions in our youth is associated with things as insignificant as finding a sexual mate, which is just more repetition of the same story: mating to produce children, and those children when they grow up will also search for a mate to have other children, and so on. That is our fate as animals, our fate as small cogs of nature. But some day in the past, human beings dreamed about a better end and greater meaning to our lives, something beyond the simple laws which regulate the behaviour of plants and animals. Our greatest intellectual heroes dreamed about a man devoted to the arts, religion, philosophy and humanities, science; they dreamed about a man who could gaze nature and say "Eh! You are not above me. I can reach the highest summits of your forces because I can understand how you work". But the dream is over, and now we realize the true essence of our species: Indeed, all those efforts to dominate nature through science are, as noted by Nietzsche, a will to power; an attempt to obtain a higher status, become rich and famous, to get the best women or men in order to mate with them and to produce the most beautiful children of whom they can be proud. We have got a science whose main end lies in getting a salary and growing fat with our mates and children, or alone, in a society with plenty of technology to make our lives more comfortable. How close to animals we are!

Science is not the same thing as religion; I have a very clear opinion about it. However, science and religion and all human affairs, apart from eating, drinking, mating, and other biological tasks, share something in common. They share the will of human beings to surpass our limited condition of simple matter, simple animals. We want to be more than dogs or cats. We want to be more than the inert mountains and rivers or the fixed stars in the dark celestial shell. A Faustian spirit, in Spengler's terms, a poetic impulse of human beings to go beyond the limits that nature has placed on them. Nonetheless, death is always present among the individuals of our society so we cannot forget that we die like dogs or cats, and that we are simple mortal dust, like the matter of the stars. And even the strongest delusions about immortality turn to disillusion when time erodes cultural possibilities. Old polytheistic religions fall, monotheistic religions have been crumbling in the enlightened western culture for the last two centuries, and now it is time for science, the last of the dreams the greatest manifestation of our culture, to fall.

I wonder what humans' dreams of the future will be in a few centuries' time. If I were a good prophet, I would be able to predict this, but I am not. I can only guess that the new humanity which will come after us will be quite different. It will behave very differently in cultural terms, though similarly in the biological aspect, of course. That which gives sense to our lives will be a historical curiosity for people in the future. Possibly, people will not visit fine art museums to look at paintings, and they will not take pictures like tourists. Possibly, the members of the new humanity will not go to concerts, conferences, cinemas, etc. Possibly, society will not invest great efforts in building sophisticated devices with which to do science. In the same way that contemporary individuals do not find too much meaning in the typical religiosity of Middle Age Europe and we do not build impressive cathedrals (with rare exceptions like the "Sagrada Familia" in Barcelona), future humanity will not have too much interest in building big telescopes and particle accelerator or toys similar to those with which the scientists play nowadays. Indeed, as pointed out by Binggeli (2006), the great and expensive telescopes or particle accelerators are comparable to the cathedrals of the Middle Age as expressions of power and prestige. Each epoch has different entertainments, or ways of diverting our thoughts from our mortality and our finite situation. But which will be the new highest expressions of culture for the new humanity? I do not know.

With regard to science, I guess that the most important parts of our knowledge will not be lost, though most of the rubbish published nowadays with plenty of details about unimportant questions is likely to be lost, rather as happened with the classical culture of Greek and Rome. I guess the new humanity will preserve the accumulated scientific knowledge in some kind of conservatories of knowledge, as happens with classical music nowadays. Some scholars, not many, might be in charge of monasteries of knowledge, where cloistered people might preserve the documents and computer data, and they will keep alive the tradition of teaching scientific topics to some pupils, but this will not affect most of the society, which will have other more important things to worry about. Of course, this is a speculation. I might be wrong.

This intuition about the future of our culture is also considered in Hermann Hesse's *The Glass Bead Game*, a novel which takes place at an unspecified date, centuries into the future. The setting is a fictional province of central Europe called Castalia, reserved by political decision for the life of the mind; technology and economics are kept to a strict minimum. The aim of that place is to preserve the spirituality inherent to the culture of past epochs which would otherwise be sunk in the midst of economic targets and technological developments, in a society embedded in a materialist/pragmatic life. Castalia is home to an austere order of intellectuals, of people who renounce the worldly life, marriage, property, family, etc., and stay in the institution with a twofold mission: to run a boarding school for boys, and to nurture and play the Glass Bead Game, whose exact nature remains elusive and whose devotees occupy a special school within Castalia. The rules of the game are only alluded to, and are so sophisticated that they are not easy to imagine. Playing the game well requires years of hard study of music, mathematics, and cultural history. Essentially, the game is an abstract synthesis of all arts and sciences, representing the aesthetic face of the totality of life in the form of rhythmic processes. It proceeds by players making deep connections between seemingly unrelated topics.

The story of *The Glass Bead Game*, or any other utopia of the far future, is just a speculation, a fantasy about the future. Indeed, nobody knows how the future will be, but there are good reasons to think that humanity is already at the beginning of a new Dark Age of culture, and there are also some reasons, or at least hope, for believing that some of our cultural treasure will remain for the future. The glass bead game is only symbolic of the new shapes that culture may adopt in the future, whatever it may be. The role that pure sciences (apart from technological applications) will occupy in the culture of the new humanity is unknown now. Let us hope it will retain our tradition of understanding how nature behaves, but in a different way from what we have known up until now. Science is dying! Let's wait and see if future generations keep the best of it in their minds. Long life to science in our memory!

8.1   Some recommendations for future politics in science

The fate of our civilization cannot be changed. Whatever has to happen, will happen. Societies develop their cultures, and they grow, reproduce and die; and its control is beyond our strength as individuals. Do not interpret this book as an attempt to convince society that science must vanish. No, that is not my intention and anyway it would be an impossible mission. I have only described what I see (from my subjective point of view, of course), not what I would like to see. Indeed, my hopes are for the dawn of a new enlightenment rather than the twilight of science.

We cannot create a programme to keep culture alive. Of course, we can give more money to those individuals who promise the moon and stars in the creation of some new cultural paradigm, but that is just an economic transaction, just market activity, nothing to do with real culture, which emerges when it has to emerge, not when funds are available (except in architecture or other fields which depend strongly on funding). Therefore, in my opinion, there is nothing the politics of science can do to stop the decline of science, and there is nothing it can do to promote a new form of culture except wait for its spontaneous emergence. I also realize that any advice will fall in deaf ears among present-day administrators. Nonetheless, in view of the present-day panorama of science that I have described in these pages, I will make some recommendations; they might be useful for future generations.

Perhaps science cannot be improved, but at least a fairer distribution of funds and resources could be reached with an appropriate political stance. Anyway, this book is not a pamphlet with political or sectarian ambitions. I am not interested in convincing anyone of any claim. I do not think that this text is useful for supporting the claims of trade unions, demanding rights for science workers. It is not helpful to claim some "right", because the present problems of scientific research will not be solved with the increase of bureaucracy; they would simply worsen.

We have too many people doing science, and few of them who have the power do not want to be mixed up with the vast hordes of creators. It is all a question of the will to power. Therefore, they prefer to maintain the barriers which separate the good science from the bad science, with all the consequent biases introduced in this selection: Lots of mediocre misguided ideas are published in important journals, whereas many bad papers and a few genuine ideas are left aside unpublished or published in minor publications which almost nobody reads. My suggestion would be to allow the important media to publish all kind of contributions, at least in an electronic version on the internet, which is cheaper. For instance, arXiv.org could stop the filtering of papers and allow the posting of all kinds of publications related to the scientific topics they cover, even papers from amateurs. Some scientists think this would be terrible because important knowledge would be drowned by huge numbers of papers by amateurs, but this is not true: It can be shown that before 2004, the date at which the filtering of papers was introduced on arXiv.org, the number of papers produced by crank amateurs was a small portion of the total; they do not disturb the flow of information produced by professional scientists, and they introduce some fun touching on questions which are not usually dealt with by paid researchers. For major journals, instead of a peer review system to ensure quality, I would suggest that anybody with a Ph.D. in any scientific discipline may publish; something similar is also proposed by Gillies (2008, ch. 9). It might be supposed that a Doctor of Philosophy already has sufficient maturity to know what he is doing, although the journals could ask experts to recommend revisions of the papers, but only to improve them with constructive criticism, not to reject them unless the author decides to withdraw the paper; or they could allow experts to post their comments on the published papers (Ietto-Gillies, 2008).

In my opinion, funding and access to instruments should also not to be distributed according to the number of publications, citations, impact factors, etc. Quality cannot be measured with any of these quantitative parameters. Committees evaluating the quality of past, present or future researches favour the mediocracies and big-science projects rather than the ideas of a minority, so they are not fair either. Rather than scientometrics or commissioning, I suggest an equitable distribution of funds and resources among experts, once they have demonstrated that they are able to do research, that is once they have got a Ph.D. For instance, in my speciality of astrophysics, I suggest that telescope time be distributed equitably among all the observational astrophysicists and when some researcher does not need their allotted time on the telescope, it can be given to another researcher. Normally, the demand of telescope time is higher than what is on offer, and the most powerful teams attempt to obtain most of the telescope time, arguing that their science is better. However, I think the researchers should keep their observations within their available time. For big science projects, many researchers might pool their telescope time. And when the need for telescope time is greater than the time available to a researcher, they might be included in a queue of researchers, to obtain some free time on the telescope when it is not being used by other researchers.

The assigning of positions, either Ph.D. grants or postdocs or permanent positions, is more difficult to solve, given that there are more applications than positions. This should be addressed with an appropriate education system, and a new direction in the job of researcher. Indeed, many of the present-day researchers do not have a true vocation for science and natural philosophy; they are mostly technocrats who might be redirected to jobs as engineers or computer experts or something similar. A reduction in the salaries of researchers would also reduce very significantly the number of people who want to do research; perhaps, the State could offer a free house and food and a small salary for other expenses, to encourage an austere way of life, dedicated to study and thinking rather than linking business and conference-tourism with science. Once this regulation is done, I am sure that the number of people who would want to do research would be much lower, and with a higher passion for science; there would be a place for all of them without need of a competitive selection system. There would also be fewer papers and conferences, less knowledge in which to be drowned, which would also be a positive outcome.

A world of science without reports, committees and referees would have another important advantage: saving money and the time of the researchers. At present, most senior researchers spent a high portion of their research time in writing reports, and reading reports of other colleagues to evaluate them. So, without peer reviewing, there would be a greater dedication to doing research. And it is also possible that scientists will dare to do research on more innovative high-risk ideas. Loeb (2010) recommends that people invest further research time in high-risk ideas, but this can only be a reality if the systems of evaluation change or disappear.

If anybody does not like these suggestions, I could suggest another alternative to maintain justice. We could apply a system of compensations and penalties according to the biases of the system. We would let the system run as it is now and, when we realized that some decision was unfair, we could correct it by punishing the people responsible for the bad decision and rewarding the scientist who suffered the consequences of that decision. For instance, let us suppose that a researcher is unable to publish a paper in a high-profile journal, so he/she has to publish in minor journals or not at all. They would get a small number of citations, would not be able to get a position in research centre, given the non-competitive CV, and would not get grants to go to conferences to announce their results. They might even have to leave research and dedicate their time to other things. Let us suppose that after many years or decades, the results of this researcher are discovered to be of high importance in their field, innovative, and a revolutionary landmark in that subject. This is not very usual, but it might happen, maybe with $1/1000$ probability or less. In that case, if it is not too late, that is, if the researcher is still alive, they should be given a prize, but an honorary prize is not enough, they should also be compensated for the money that they could not earn with many years of research. On the other hand, all those people involved in the negative evaluation of that work should be announced publicly if anonymous and punished; it is fair, in order to compensate the damage in the career of the scientist who was rejected. The referees of major journals who rejected their papers should be punished economically or even fired if their unfairness is demonstrated. The institutions which refused to give the researcher a position could be punished financially. Some institutions should be even closed if it is shown than in recent years they have invested their money in following the wrong lines of research, damage the innovative research. For instance, imagine that within some decades it is shown that ideas about dark matter and dark energy in present-day cosmology are totally wrong while some other heterodox idea whose researchers suffered obstruction was correct. In such a case, the many tens of research institutes which are now dedicating most of their funds to standard cosmology and the search of dark matter or similar topics, while closing their doors to scientists with other, alternative views, should be closed. It is fair. If the lives of scientists with better ideas were shipwrecked in favour of a cast of mediocre scientists, it is fair that the working lives of those mediocre scientists be now cut short and the better scientists elevated to a higher position. If a scientist feel that he has the right to cut short the scientific life of a person by judging the quality of their science, it is fair that the system cuts short the scientific life of that judge if he has made an unfair decision. Maybe some generations would go by before the mistake in a line of research is discovered; in such a case, the researcher's heirs for two or three generations might receive their prizes or penalties: economical compensation for the heirs of the scientist who was treated unfairly, and financial penalties for the heirs of the scientists who obstructed the research of the first one unfairly, in the sense that the inherited properties would be confiscated. It is fair because the goods of the obstructive scientists were more than they deserved because of their jobs. Possibly, we may consider these penalties to be very hard. The sanctioned researchers would reply that their lines of research might be wrong but science is not a straight path and some misconceptions are part of the game. Certainly, science is not a straight path: hypotheses that now we consider correct may be shown to be wrong tomorrow, and vice verse; but this should also be borne in mind by the unreasonable referee or committee before rejecting a paper containing innovative ideas. If somebody is pretty sure that an innovative idea is wrong, in order to reject it they should be able to bet his possessions on it being so; otherwise, it would mean that they are not so sure. Life is hard for the tough guys. He who lives by the sword dies by the sword. Certainly, it would be a tougher game. I do not like the option of this paragraph, it would be a monster of bureaucracy feeding the lawyers and annoying the courts. It would be easier for everybody if rather than adopting the previous recommendation, there were no committees and no referees separating good science from bad science.

Natural science should be an open space belonging to all those restless minds who want to think about how nature is, rather than a feud between a few owners who want to defend their status. It is important that young people, younger than forty years old, gain higher status in the system, because most of the novel ideas and innovations are likely to come from them, not from the establishment. Of course, older scientists will always defend their status, based on their experience and this has great value too, but for the game of research in science, innovation is more important than experience. Certainly, in the age of the twilight of science, there may be not too much left to innovate but, if there is some space, it should be most of it left for young people. As I said, referring to Ph.D. students, rather than loading young researchers with monotonous work, it should be in the hands of those whose creativity is exhausted, those aged, reputed experts who will produce nothing but copies of what they have always produced. Naturally, older scientist will rebel and will not want to do the hard tasks. Then, the best solution is that everybody works on their own ideas and nobody is the slave of anybody. If you have an idea, work at it yourself. Hiring a postdoc to do a predetermined task should be forbidden; postdocs should be free to choose their topic of research from among their own ideas. Big science should emerge only from big collaborations among equal-status scientists, not from the hierarchical order of bosses (old people who direct the use of state funds for science) and slaves (students and postdocs who waste their talents in doing routine tasks for other people without ideas).

The problem is what to do with the scientists once they get older. Nowadays they tend to occupy administrative positions in science, in which they are very comfortable, but I think the number of people dedicated to bureaucratic control and administration should be much reduced. Maybe we could consider the career of a researcher as being limited in time, like a sportsman or a ballet dancer, and they should later move to other positions, such as teaching in the university or even away from science. And, of course, they could do research if they were not yet bored with it, but not from a privileged position. Any place where they did not disturb the young scientists should be ok. I do not recommend a complex and inefficient structure of administrators reading reports and controlling the flow of science as at present; rather, in agreement with Gillies (2008), I think the figure of scientist-administrator should vanish or should be very much reduced to tasks that are carried out by administrative secretaries. Honestly, I think old scientists should get early retirement or they should change their job once they do not feel strong to do research with their own hands but instead exploit young people. But we all know that this will not happen, because the power is in the hands of old people in our society. Indeed, I may be myself adopting a cynical position because I do not plan to leave research now I am 41 years old (at the beginning of 2012), and I am acquiring more power in science, once I have a permanent staff position as researcher. But if somebody asks my opinion, I will always say that I am in favour of a reduction in the power of old people in science, in favour of young people, provided that the individuals of the new generations are well prepared as scientists, which is something that needs to be discussed too, given the sharp decline of quality in graduate students at present.

Sylos Labini & Zapperi (2010) claimed that there is no future for our societies without research, perhaps with the intention to encourage the Italian state to invest further money in research. I can however easily imagine a future without research. Indeed, this will be the most likely situation in the future: States will not give priority to research over other things in society. Building hospitals is more important than doing research, even in medicine or pharmacy.

Science is already very advanced in all its fields. In order to avoid a Sisyphean science, we must put an end to science, now or soon, and after this end we must consider the mission of science to be accomplished. The eternal progress of our science cannot be maintained. In technological applications, progress will last longer, but there will also be some point at which no further development is possible. Governments should bear this in mind, and they should plan, in the long term, a slow, continuous reduction of staff, first to stop the exponential increase of the last few decades, and later to decrease the absolute number of researchers and research centres and their associated expenses. This should be done slowly to avoid dramatic situations for the people who have already spent most of their lives in research.

Over the last century, we have got used to the idea that good science is produced only with huge amounts of money, a capitalist idea on which US administrators are very keen. But nations should realize that this competition among countries to get the best marks in science by means of higher investments is absurd and it leads to a science which is more and more expensive but producing less and less interesting returns. In my opinion, the decline of science cannot be halted by asking for further money. Quite the opposite: the more money is invested, the more the system becomes a mafia. No, the decline of science is unavoidable. The golden age of science will never come again. But at least we could try to preserve the competitive spirit of science, in which the best intelligences can produce smart solutions to various problems. I bet on creativity, I bet on the intellectual emotions, for the high emotions of the scientific intelligence, and I condemn the science made with very expensive devices in which intelligence is replaced by the manager as administrator and by the ability to get money from the state. Therefore, I recommend for the future administration of science that states avoid investments in mega-huge experimental projects. Rather, with a much reduced budget, money should be invested in people rather than instruments, people who are able to think, to produce ideas. We must not forget that, above all, the value of science is in the ideas. Experiments are also important for natural science. Physics is not mathematics, but an empirical science. But that empirical science should be constrained by a limited budget. Of course, it is very easy to imagine the fantastic experiments which could be created with unlimited budgets; maybe we could create a particle accelerator of the size of the earth; but this has no merit. Thinking about new ideas with low-budget experiments has more merit, and this should be favoured. We must also remember that ideas are produced by individuals, not by big institutions, so favouring an individualistic science, against the modern trends of large teams, would save something of the true spirit of science: intelligence. Many scientists might possibly complain about this political move and say: "With a low budget, we cannot create innovative science". And the answer to them should be firm as well: "If you cannot produce new ideas or new analyses of available data in science, and your only idea is to ask for more money for a device more expensive than the previous one, then leave research".

A drastic reduction of funds for conferences and travels expenses would also be advisable. We are living a time with very serious environmental problems, such as global warming, and by the time the present advice is taken into account, not be before 2040 or 2050, the environmental problems will be even more severe. It does not matter whether scientists waste time and money doing bad science, but the environmental problems are much more important. Therefore, it would make sense for the new priests of knowledge to be associated with a new moral order giving a good example with their behaviour. It would be nice if scientists travelled only very occasionally (once every few years) and avoided the pollution produced by planes, prompted by the organization of so many conferences as are held nowadays.

For the new humanity, science will occupy a nostalgic position in society, as something belonging to the golden epochs of the past. Society should be educated in the classical values of science, both with a good education in the school and universities, and with information in the mass media. Nowadays, the media prefer to talk about the latest discoveries of science, the latest paradigms and speculations. But in a science with fewer and fewer returns, or returns which are interesting but not reliable (for instance, in my field of cosmology), people get confused and do not learn as much from that propaganda. A redirection of the dissemination of scientific information in terms of well established classical concepts of science is preferable for the general public. Science is a treasure of our culture. We must look after it, and transmit it to the grandchildren of our grandchildren, and so on, for centuries and centuries.

8.2 Twilight

The spirit of science is dying, but it is mature enough to die with dignity. Our fate as human beings who have known the high peaks of knowledge is a "Götterdämmerung", a twilight of the gods, as in Wagner's opera. No, we are not gods and we have never been gods, but we have dreamt about the immortality of knowledge in the whole universe. We have dreamt about absolute power over nature. Those delusions of grandeur made us great. They have created light in the dark spaces of our existence. Also, it is not the end of the world in an apocalyptical way. The death of our science, of our whole culture indeed, is coming slowly. We are at the beginning of a transitional period which will transform our present culture into a new one.

Within some centuries, humankind will contemplate the ruins of our civilization in the same way that romantic poets contemplated the ruins of past cultures with the sweet allure of decay, death and destruction. Science will prompt an emotional response again and will be revived. A new dawn after the twilight? Who knows? In any case, the eternal values never die, and Truth will always be Truth.

REFERENCES

Akahira, M., & Mizubayashi, H., 2009, "Reply letter", *Physics Today*, July 2009, p. 11

Andreas-Salomé, L., 1894, *Friedrich Nietzsche in seinen Werke*. Translated into English in: 1988, *Nietzsche*, Black Swan Books, Redding Ridge (CT, USA)

Arana, J., 2001, *Materia, Universo, Vida*, Tecnos, Madrid

Arana, J., 2003, "Sobre la situación actual de la Universidad. Problemas y soluciones", *LOGOS. Anales del Seminario de Metafísica*, 36, pp. 41-48

Arana, J., 2012, *Los sótanos del Universo*, Biblioteca Nueva, Madrid

Arp, H. C., 2008, "Scientific and political elites in Western democracies", in: *Against the Tide. A Critical Review by Scientists of How Physics & Astronomy Get Done*, M. López Corredoira, C. Castro Perelman, Eds., Universal Publishers, Boca Raton (Florida), pp. 117-128

Armstrong, J. S., 1997, "Peer review for journals: evidence on quality control, fairness, and innovation", *Science and Engineering Ethics*, 3, pp. 63-84

Asimov, I., 1959, *Breakthroughs in Science*, Scholastic Magazines Inc., New York

Battaner, E., 2006, *Un físico en la calle*, Universidad de Granada, Granada (Spain)

Battaner, E., 2010, *El astrónomo y el templario*, Nabla ediciones, Alella (Barcelona)

Bauer, H. H., 2008, "Ethics in Science", in: *Against the Tide. A Critical Review by Scientists of How Physics & Astronomy Get Done*, M. López Corredoira, C. Castro Perelman, Eds., Universal Publishers, Boca Raton (Florida), pp. 239-255

Bauer, H. H., 2012, *Dogmatism in Science and Medicine. How Dominant Theories Monopolize Research and Stifle the Search for Truth*, McFarland, Jefferson (NC, US)

Berk, H. L., et al., 2008, "Scientists protest professor's dismissal", *Physics Today*, December 2008, pp. 10-11

Berk, H. L., & Fisch, M., 2009, "Reply letter", *Physics Today*, July 2009, p. 11

Binggeli, B., 2006, *Primum Mobile. Dantes Jenseitsreise und die moderne Kosmologie*, Ammann Verlag & Co., Zurich

Braben, D. W., 2004, *Pioneering Research. A Risk Worth Taking*, John Wiley, Hoboken (NJ)

Brezis, E. S., 2007, "Focal randomization: an optional mechanism for the evaluation of R&D projects", *Science and Public Policy*, 34, pp. 691-698

Brockman, J., 1995, *The Third Culture*, Simon & Schuster, New York

Brown, N. O., 1966, *Love's body*, Random House, New York

Calvin, M., 1956, "Chemical Evolution and the Origin of Life", *American Scientist*, 44, pp. 248-263

Campanario, J. M., & Martin, B., 2004, "Challenging dominant Physics paradigms", *Journal of Scientific Exploration*, 18(3), pp. 421-438

Castro Perelman, C., 2008a, "My Struggle with Ginsparg (arXiv.org) and the Road to Cyberia: a Scientific-Gulag in Cyberspace", in: *Against the Tide. A Critical Review by Scientists of How Physics & Astronomy Get Done*, M. López Corredoira, C. Castro Perelman, Eds., Universal Publishers, Boca Raton (Florida), pp. 59-76

Castro Perelman, C., 2008b, "Is Ethics in Science an Oxymoron?", in: *Against the Tide. A Critical Review by Scientists of How Physics & Astronomy Get Done*, M. López Corredoira, C. Castro Perelman, Eds., Universal Publishers, Boca Raton (Florida), pp. 257-265

Cho, T., 2009, "Fired Tsukuba professor's defence", *Physics Today*, July 2009, p. 10

Cho, T., et al., 2006, "Observation and Control of Transverse Energy-Transport Barrier due to the Formation of an Energetic-Electron Layer with Sheared $E \times B$ Flow", *Phys. Rev. Lett.*, 97, 055001

Conan Doyle, A., 1887, *A Study in Scarlet*, Ward Lock & Co., London

Conan Doyle, A., 1892, "A Scandal in Bohemia", in: *The Adventures of Sherlock Holmes*, George Newnes Ltd., London

Dawkins, R., 1976, *The Selfish Gene*, Oxford University Press, Oxford

Descartes, R., *Regulae ad directionem ingenii*. Translated into English in: 2000, *Rules for the Direction of the Mind*, Bobbs-Merrill Co., Indianapolis (Indiana) [Translations into English throughout this book made by myself]

Dietrich, J. P., 2008a, "The Importance of Being First: Position Dependent Citation Rates on arXiv:astro-ph", *Publ. Astron. Soc. Pacific*, 120, pp. 224-228

Dietrich, J. P., 2008b, "Disentangling Visibility and Self-Promotion Bias in the arXiv:astro-ph Positional Citation Effect", *Publ. Astron. Soc. Pacific*, 120, pp. 801-804

Diggins, F., 2003, *The true history of the discovery of penicillin by Alexander Fleming*, Biomedical Scientist, March 2003, Institute of Biomedical Sciences, London

Dolsenhe, O., 2011, *A Critique of Science: How Incoherent Leaders Purged Metaphysics of Mind and God*, Lulu, Raleigh (NC, USA)

Elger, C. E., et al., 2004, , "Das Manifest", *Gehirn & Geist*, 6-04, pp. 30-31

England, P., Molnar, P., & Righter, F., 2007, "John Perry's neglected critique of Kelvin's age for the Earth: A missed opportunity in geodynamics", *GSA Today*, 17 (1), pp. 4-9

Feyerabend, P. K. 1970, "Philosophy of Science: A Subject with a Great Past", *Historical and Philosophical Perspective of Science. Minnesota Studies in the Philosophy of Science*, vol. 5, R. H. Stuewer, ed., Minneapolis, Univ. of Minnesota Press

Feyerabend, P. K., 1975, *Against Method: Outline of an Anarchistic Theory of Knowledge*, Humanities Press, Atlantic Highlands (NJ, USA)

Feynman, R. P., 1974, "Cargo Cult Science", *Engineering and Science*, 37, 7

Fox, S. W., 1956, "The Evolution of Protein Molecules and Thermal Synthesis of Biochemical Substances", *American Scientist*, 44, pp. 347-359

Frey, B. S., 2003, "Publishing as prostitution? Choosing between one's own ideas and academic success", *Public Choice*, 116, pp. 2005-223

Fukuyama, F., 1989, "The End of History?", *The National Interest*, 16, pp. 3-18

Fukuyama, F., 1992, *The End of History and the Last Man*, Free Press, New York

Gal, O., 2002, *Meanest Foundations and Nobler Superstructures. Hooke, Newton and the "Compounding of the Celestial Motions of the Planets"*, Kluwer, Dordrecht

García Márquez, G., 1981, *Crónica de una muerte anunciada*, Editorial Diana, México. Translated into English in: 1983, *Chronicle of a Death Foretold*, Alfred A. Knopf, New York

Gascoigne, R., 1992, "The Historical Demography of the Scientific Community, 1450-1900", *Social Studies of Science*, 22(3), pp. 545-573

Gibbon, E., 1776-1789, *The History of the Decline and Fall of the Roman Empire*, Strahan & Cadell, London

Giddens, A., 2006, *Sociology* (5th edition), Polity Press, Cambridge (U.K.)

Gillies, D., 2008, *How Should Research be Organised?*, College Publications, London

Godlee, F., Gale, C. R., & Martyn, C. N., 1998, "The effect on the quality of peer review of blinding reviewers and asking them to sign their reports: a randomised controlled trial", *J. Am. Med. Assoc.*, 280, pp. 237-240

Gómez Dávila, N., 2007 (post.), *Escolios escogidos*, Ed. Los Papeles del Sitio, Sevilla (Spain)

Gould, S. J., 1989, *Wonderful Life: The Burgess Shale and the Nature of History*, W. W. Norton, New York

Habing, H. J., 2009, "Citation counts: trick or treat? Commentary on: Kessler M. F., Steinz J. A., Anderegg M. E., et al., 1996, A&A, 315, L27", *Astronomy & Astrophysics*, 500, pp. 499-500

Hawking, S., 1988, *A Brief History of Time*, Bantam Books, New York

Heath, T., 1913, *Aristarchus of Samos, the ancient Copernicus; a history of Greek astronomy to Aristarchus, together with Aristarchus's Treatise on the sizes and distances of the sun and moon: a new Greek text with translation and notes*, Oxford University Press, London

Herndon, M., 2008, "Basic cause of current corruption in American science", in: *Against the Tide. A Critical Review by Scientists of How Physics & Astronomy Get Done*, M. López Corredoira, C. Castro Perelman, Eds., Universal Publishers, Boca Raton (Florida), pp. 77-85

Hesse, H., 1943, *Das Glasperlenspiel*. Translated into English in: 1969, *The Glass Bead Game*, Holt, Rinehart & Winston, New York

Hessen, B. M., 1931, "The Social Roots of the Newton's Principia", in: *Science at the Cross Roads*, Kniga Ltd., London, pp. 151-212

Hilts, P. J., 1992, "The Science Mob: The David Baltimore Case – And Its Lessons", *The New Republic*, 25, pp. 28-31

Hogan, J. P., 2004, *Kicking the sacred cow*, Baen-Riverdale, New York

Hogg, D. W., 2009, "Is cosmology just a plausibility argument?", http://arxiv.org/abs/0910.3374

Horgan, J., 1996, *The end of science. Facing the Limits of knowledge in the twilight of the scientific age*, Addison Wesley, Reading (Massachusets)

Horkheimer, M., & Adorno, T. W., 1947, *Dialektik der Aufklärung*, Querido Verlag, Ámsterdam. Translated into English in: 2002, *Dialectic of Enlightenment*, Stanford University Press, Stanford

Horrobin, D. F., 1990, "The philosophical basis of peer review and the suppresion of innovation", *J. Am. Med. Assoc.*, 263, pp. 1438-1441

Huxley, T. H., 1895, "The Scientific Aspects of Positivism", in: *Lay Sermons, Addresses and Reviews*, London

Ietto-Gillies, G., 2008, "A XXI-century alternative to XX-century peer review", *Real-world economics review*, 45, pp. 10-22

Iradier, M., 2009, *La ciencia en coordenadas*, Hurqualya, Cádiz (Spain)

Janis, I. L., 1972, *Groupthink: psychological studies of policy decision and fiascos*, Houghton Mifflin, Boston MA

Kennefick, D., 2009, "Einstein Versus the Physical Review", *Physics Today*, 58, 43

Knuteson, B., 2007, "A Quantitative Measure of Experimental Scientific Merit", http://arxiv.org/abs/0712.3572

Knuteson, B., 2009, "A Quantitative Measure of Theoretical Scientific Merit", http://arxiv.org/abs/0909.2361

Krisis (collective), 1999, "Manifest gegen die Arbeit", http://www.krisis.org/1999/manifest-gegen-die-arbeit

Kuhn, T. S., 1962, *The Structure of Scientific Revolutions*, University of Chicago Press, Chicago

Kundt, W., 2008, "The Gold effect: odyssey of scientific research", in: *Against the Tide. A Critical Review by Scientists of How Physics & Astronomy Get Done*, M. López Corredoira, C. Castro Perelman, Eds., Universal Publishers, Boca Raton (Florida), pp. 37-58

Lakota, B., & Navarro, I., 2006, "El regreso de Freud", *XLsemanal* (966), pp. 24-34

Levere, T. H., 2001, *Transforming Matter – A History of Chemistry for Alchemy to the Buckyball*, The Johns Hopkins University Press, Baltimore, Maryland

Llinás, R., 2001, *I of the Vortex: From Neurons to Self*, MIT Press, Cambridge, MA (US)

Loeb, A., 2010, "Taking 'The Road Not Taken': On the Benefits of Diversifying Your Academic Portfolio", http://arxiv.org/abs/1008.1586

Longair, M. S., 2001, "The Technology of Cosmology", in: *Historical Development of Modern Cosmology* (ASP Conf. Ser. 252), V. J. Martínez, V. Trimble, M. J. Pons-Bordería, eds., The Astronomical Society of the Pacific, S. Francisco, p. 55

López Corredoira, M., 2005, *Somos fragmentos de Naturaleza arrastrados por sus leyes*, Vision Net, Madrid

López Corredoira, M., 2008a, "What do astrophysics and the world's oldest profession have in common?", in: *Against the Tide. A Critical Review by Scientists of How Physics & Astronomy Get Done*, M.

López Corredoira, C. Castro Perelman, Eds., Universal Publishers, Boca Raton (Florida), pp. 145-177

López Corredoira, M., 2008b, "What is Research?", in: *Against the Tide. A Critical Review by Scientists of How Physics & Astronomy Get Done*, M. López Corredoira, C. Castro Perelman, Eds., Universal Publishers, Boca Raton (Florida), pp. 219-225

López Corredoira, M., 2009a, "Sociology of Modern Cosmology", in: J. A. Rubiño Martín, J. A. Belmonte, F. Prada, & A. Alberdi, Eds., *Cosmology across Cultures* (ASP Conf. Ser. 409), ASP, S. Francisco, pp. 66-73

López Corredoira, M., 2009b, "¡Viva el desempleo!, ¡abajo el consumo!", *El Manifiesto contra la muerte del espíritu y la tierra*, 11, pp. 67-89

Lovelock, J., 1979, *Gaia: a new look at life on Herat*, Oxford University Press, Oxford

Luminet, J.-P., 2008, "Is Science Nearing Its Limits? Summarizing Dialogue", http://arxiv.org/abs/0804.1504

Mackenzie, D., 2006, "Breakthrough of the year. The Poincaré Conjecture—Proved", *Science*, 314, pp. 1848-1849

Maddox, J., 1993, "Is the Principia Publishable Now?", *Nature*, Aug. 5 1995, p. 385

Mahoney, M., 1977, "Publication prejudices: an experimental study of confirmatory bias in the peer review system", *Cognitive Therapy and Research*, 1, pp. 161-175

Mann, T., 1924, *Der Zauberberg*. Translated into English in: 1927, *The Magic Mountain*, Secker and Warburg, London [Translations into English throughout this book made by myself]

Martin, B., 1997, *Suppression Stories*, Fund for Intellectual Dissent, Wollongong (Australia)

Martin, B., 2010, "How to Attack a Scientific Theory and Get Away with It (Usually): The Attempt to Destroy an Origin-of-AIDS Hypothesis", *Science as Culture*, 19(2), pp. 215–239

Martínez, V. J., & Trimble, V., 2009, "Cosmologists in the Dark", in: J. A. Rubiño Martín, J. A. Belmonte, F. Prada, & A. Alberdi, Eds., *Cosmology across Cultures* (ASP Conf. Ser. 409), ASP, S. Francisco, pp. 47-56

Mizubayashi, H., 2009, "University of Tsukuba defends professor's dismissal", *Physics Today*, February 2009, p. 12

Merton, R. K., 1968, "The Matthew Effect in Science", *Science*, 159(3810), pp. 56-63

Miller, A. I., 2001, *Einstein, Picasso*, Basic Books, New York

Morris, D., 1967, *The Naked Ape: A Zoologist's Study of the Human Animal*, Jonathan Cape, London

Mulkay, M. J., 1976, "Norms and ideology in science", *Social Science Information*, 15, pp. 637-656

Nietzsche, F., 1880, *Der Wanderer und sein Schatten*, Schmeitzner, Chemnitz. Translated into English as *The Wanderer and his Shadow* [Translations into English throughout this book made by myself]

Nietzsche, F., 1881, *Morgenröthe*, Chemnitz. Translated into English in: 2007, *The Dawn of Day*, Dover, New York [Translations into English throughout this book made by myself]

Nietzsche, F., 1901 (posth.), *Der Wille zu Macht*. Translated into English in: 1968, *The Will to Power: In Science, Nature, Society and Art*, Random House, New York [Translations into English throughout this book made by myself]

Oblomoff (collective), 2009, *Un futur sans avenir. Pourquoi il ne faut pas sauver la recherche scientifique*, L'échappée, Paris

Ordóñez, J., Navarro, V., & Sánchez Ron, J. M., 2004, *Historia de la ciencia*, Espasa Calpe, Madrid

Ortega y Gasset, J., 1929, *La rebelión de las masas*. Translated into English in: 1994, *The Revolt of the Masses*, W. W. Norton, New York

Patterson, C. C., 1956, "Age of meteorites and the earth", *Geochimica et Cosmochimica Acta*, 10(4), pp. 230-237

Peres, A., 2002, "Karl Popper and the Copenhagen interpretation", *Studies in History and Philosophy of Modern Physics*, 33, pp. 23-34

Peters, D. P., & Ceci, S. J., 1982, "Peer review practices of psychological journals: the fate of published articles, submitted again", *Behavioural and Brain Sciences*, 5, pp. 187-195

Racker, E., 1989, "A view of misconduct in Science", *Nature*, 339, pp. 91-93

Rand, A., 1943, *The Fountainhead*, Bobbs-Merrill, Indianapolis (US)

Ritter, M., 2010, "Russian mathematician rejects $1 million prize", AP on Yahoo!News on 2010-07-01

Rosen, E., 1995, *Copernicus and his Successors*, Hambledon Press, London

Rothwell, P. M., & Martyn, C. N., 2000, "Reproducibility of peer review in clinical neuroscience: is agreement between reviewers any greater than would be expected by chance alone?", *Brain*, 123, pp. 1964-1969

Russell, B., 1946, *A History of Western Philosophy*, George Allen & Unwin Ltd., London

Salpeter, E. E., 2005, "Fallacies in astronomy and medicine", *Rep. Prog. Phys.*, 68, pp. 2747-2772

Schopenhauer, A., 1851, *Parerga und Paralipomena. Kleine philosophische Schriften.* Translated into English in: 2000, *Parerga and Paralipomena*, Clarendon Press, Oxford [Translations into English throughout this book made by myself]

Schrödinger, E., 1996 (post.), *Science and Humanism*, Cambridge Univ. Press, Cambridge

Schroer, B., 2011, "The holistic structure of causal quantum theory, its implementation in the Einstein-Jordan conundrum and its violation in more recent particle theories", http://arxiv.org/abs/1107.1374

Ségalat, L., 2009, *La science à bout de souffle?*, Seuil, Paris

Selve, R. Q., Ed., 1992, *The end of science. Attack and defence*, University Press of America, Lanham, Md.

Singer, C., 1931, A short history of biology, Clarendon Press, Oxford

Smolin, L., 2006, *The Trouble with Physics: The Rise of String Theory, The Fall of a Science, and What Comes Next*, Houghton Mifflin Hartcourt, Boston (Massachusets)

Soler Gil, F. J., 2008, *Lo divino y lo humano en el universo de Stephen Hawking*, Ediciones Cristiandad, Madrid

Spengler, O., 1923, *Der Untergang des Abendlandes* (2nd ed.). Translated into English in: 1991, *The Decline of the West*, Oxford UP, New York [Translations into English throughout this book made by myself]

Stahl, W. Translated into English in: 1977, *Martianus Capella and the Seven Liberal Arts, vol. 2, The Marriage of Philology and Mercury*, Columbia Univ. Pr, New York 1977

Stanek, K. Z., 2008, "How long should an astronomical paper be to increase its Impact?", http://arxiv.org/abs/0809.0692

Stent, G. S., 1969, *The Coming of the Golden Age. A view of the End of Progress*, Natural History Press, Garden City, N. Y.

Sylos Labini, F., & Zapperi, S., 2010, *Senza Ricerca non c'è futuro*, Editori Laterza, Roma

Tarver, M., 2007, "Why I am Not a Professor OR The Decline and Fall of the British University", http://www.lambdassociates.org/blog/decline.htm

Taschner, R., 2007, "Erosion von Wissenschaft", *Erwägen-Wissen-Ethik*, 18(1), pp. 58-59

Tegmark, M, Aguirre, A., Rees, M. J., & Wilczek, F., 2006, "Dimensionless constants, cosmology, and other dark matters", *Phys. Rev. D*, *73(2), 3505*

Teresi, D., 2003, *Lost Discoveries: The Ancient Roots of Modern Science*, Simon & Schuster, New York, pp. 213–214

Thurner, S., & Hanel, R., 2010, "Peer-review in a world with rational scientists: Toward selection of the average", http://arxiv.org/abs/1008.4324

Toynbee, A. J., 1935-1962, *A Study of History* (10 vols.), Oxford University Press, Oxford

Trimble, V., & Ceja, J. A., 2008, "Productivity and impact of astronomical facilities: Three years of publications and citation rates", *Astron. Nachrichten*, 329, pp. 632-647

Unamuno, M. de, 1913, *Del sentimiento trágico de la vida*. Translated into English in: 1954, *Tragic Sense of Life*, Dover, New York. [Translations into English throughout this book made by myself]

Unamuno, M. de, 1930, *San Manuel Bueno, mártir*. Translated into English in: 2009, *Saint Manuel Bueno, martyr*, Aris & Philips, Warminster (U.K.) [Translations into English throughout this book made by myself]

Unzicker, A., 2010, *Vom Urknall zum Durchknall. Die absurde Jagd nach der Weltformel*, Springer, Berlin

Van der Waerden, B. L., 1987, "The Heliocentric System in Greek, Persian and Hindu Astronomy", *Annals of the New York Academy of Sciences*, 500 (1), pp. 525–545

Van Flandern, T., 1993, "Peer Pressure and Paradigms", in: *Dark Matter, Missing Planets and New Comets*, North Atlantic Books, Berkeley

Velázquez Fernández, H., 2007, *¿Qué es la Naturaleza? Introducción filosófica a la historia de la ciencia*, Porrúa, México

Wali, K. C., 1991, *Chandra: a Biography of S. Chandrasekhar*, University of Chicago Press, Chicago

Weaver, D., Reis, M. H., Albanese, C., Costantini, F., Baltimore, D., & Imanishi-Kari, T, 1986, "Altered repertoire of endogenous immunoglobulin gene expression in transgenic mice containing a rearranged mu heavy chain gene", *Cell*, 45(2), pp. 247-259

Wenneras, C., & Wold, A., "Nepotism and sexism in peer review", *Nature*, 387, pp. 341-343

White, S. D. M., 2007, "Fundamentalist physics: why Dark Energy is bad for astronomy", *Rep. Prog. Phys.*, 70(6), pp. 883-897

Yao, W. M., Amsler, C., Asner, D., et al., 2006, "Review of Particle Physics", *Journal of Physics G: Nuclear and Particle Physics*, 33(1), pp. 1-1232

Zweig, S., 1942, *Die Welt von Gestern*, Stockholm. Translated into English in: 1964, *The World of Yesterday*, Univ. of Nebraska Press, Nebraska

CITED NAMES INDEX

CPSIA information can be obtained at www.ICGtesting.com
Printed in the USA
LVOW102102040613

336947LV00031B/1715/P